IRREFUTABLE EVIDENCE

IRREFUTABLE EVIDENCE

Adventures in the History of Forensic Science

Michael Kurland

Ivan R. Dee

Chicago 2009

www.ivanrdee.com

Library of Congress Cataloging-in-Publication Data:
Kurland, Michael.
  Irrefutable evidence / Michael Kurland.
      p. cm.
  Includes bibliographical references and index.
   ISBN 978-1-56663-803-6 (cloth : alk. paper)
   1. Evidence, Criminal—United States. 2. Evidence, Expert—United States.
3. Forensic sciences. I. Title.
   KF9660.K87 2009
    363.25—dc22   2009022386

*I dedicate this book to
The Bad Companions,
a constant source of inspiration and delight.
You know who you are.*

Contents

Acknowledgments

MY THANKS to Linda Robertson for her assistance in research and writing, and for quelling some of my more outrageous ideas. Thanks also to Bicka Barlow, Barry Scheck, William Thompson, and Sandy Zabell, who have shared their knowledge and insights with the forensic community and caused some of it to transfer to me in passing.

M. K.

Petaluma, California
July 2009

But Thomas, one of the twelve, called Didymus, was not with them when Jesus came.

The other disciples therefore said unto him, We have seen the LORD. But he said unto them, Except I shall see in his hands the print of the nails, and put my finger into the print of the nails, and thrust my hand into his side, I will not believe.

And after eight days again his disciples were within, and Thomas with them: then came Jesus, the doors being shut, and stood in the midst, and said, Peace be unto you.

Then saith he to Thomas, Reach hither thy finger, and behold my hands; and reach hither thy hand, and thrust it into my side: and be not faithless, but believing.

—*John 1:24–27*

IRREFUTABLE EVIDENCE

Introduction

> There has been murder done, and the
> murderer was a man. He was more than six
> feet high, was in the prime of life, had small
> feet for his height, wore coarse, square-toed
> boots and smoked a Trichinopoly cigar.
> —Sherlock Holmes in *A Study in Scarlet*

OVER THE PAST TWO DECADES forensic science has become a national spectator sport. A substantial majority of the reading and television-viewing public follows the exploits of both real and fictional criminalists as they go about apprehending the wicked. Several successful contemporary fiction series—among them those by P. D. James, Patricia Cornwell, Kathy Reichs, and Thomas Harris—make forensic investigation a major part of their plot. Fingerprint analysis, DNA, ballistics, blood spatter, toolmarks, body temperature, and rigor mortis are now topics of everyday conversation. There is scarcely a twelve-year-old child who isn't familiar with the comparison microscope.

The television show *CSI* (for Crime Scene Investigator) and its spinoffs, in which criminalists use the scientific method to solve crimes, have been among the top prime-time shows for the past decade. In a wave of other shows—*Bones, Cracker, Criminal Minds, Crossing Jordan, Da Vinci's Inquest, Dead Men Talking,*

3

Extreme Forensics, FBI Files, Forensic Files, NCIS, Post Mortem, Prime Suspect, Profiler, Secrets of Forensic Science, Silent Witness, and *Solved*—the forensic investigation of crime is an integral part of the story line. Autopsy close-ups and other gruesome details have replaced the car chases and gratuitous violence of yesteryear.

As one might expect from television programs, with their time limits and need for continuous action, much of the technical detail is wrong, and some of it even wildly incorrect: time intervals are condensed, science is blended skillfully into science fiction, and characters have skills they would never possess in real life. This is not a complaint—a television show's purpose is to entertain, and no one expects a show about doctors, lawyers, cops, convicts, or, for that matter, spirit mediums to be in 100 percent correlation with the real world.

When did this popular interest begin? In the early years of the nineteenth century, the rise of literacy occasioned a demand for cheap, sensational literature. And this demand was filled in part by true crime stories. Lurid newspaper accounts were often released as pamphlets featuring overly graphic covers.

James Gordon Bennett, publisher of the *New York Herald*, became the first investigative reporter when he personally examined the scene of the murder in 1836 of twenty-three-year-old Helen Jewett, one of the girls of the Palace of Passions on Thomas Street in Manhattan. Here he describes the girl's body in "classical" terms:

> Slowly I began to discover the lineaments of the corpse, as one would the beauties of a statue of marble. It was the most remarkable sight I ever beheld—I never have, and never expect to see such another. "My God," exclaimed I, "how like a statue! I can scarcely conceive that form to be a corpse." Not a vein was to be seen. The body looked as white—as full—as polished as the purest Parisian marble. The perfect figure—the exquisite limbs—the fine face—the

full arms—the beautiful bust—all—all surpassed in every respect the Venus de Medicis, according to the casts generally given of her. . . . For a few moments I was lost in admiration at this extraordinary sight—a beautiful female corpse—that surpassed the finest statue of antiquity.

This overwrought description is certainly inaccurate. In fact, after she was battered and hacked to death with an ax, the poor girl's body was set on fire.

The case of the "Beautiful Cigar Girl," Mary Rogers, served as a platform for Bennett's attack on city government. Three days after disappearing from her Manhattan home, Rogers was found floating in the Hudson River off Hoboken, New Jersey.

The recent awful violation and murder of an innocent young woman—the impenetrable mystery which surrounds that act—the apathy of the criminal judges, sitting on their own fat for a cushion bench—and the utter inefficiency of the police, are all tending fast to reduce this large city to a savage state of society—without law—without order—and without security of any kind.

The case was never solved. The known facts provided the background for several fictionalized accounts, among them J. H. Ingraham's dime novel *The Beautiful Cigar Girl or the Mysteries of Broadway* and Edgar Allan Poe's *The Mystery of Marie Rogêt*. Poe turned New York into Paris and the Hudson into the Seine. Under that guise, as he explained in a letter to a friend, he provided a rigorous analysis of the actual case and suggested several possible solutions.

In the late nineteenth century, Conan Doyle's Sherlock Holmes and R. Austin Freeman's Dr. Thorndyke blazed the fictional path toward rational crime-solving. Holmes used deduction (actually induction) with a dash of science, and Thorndyke added a heavy dose of scientific analysis. It was around this time that the first scientific crime solvers appeared in the real-life police forces of several European countries. Alphonse Bertillon, Edward Richard Henry, Hans Gross, Bernard Spilsbury, and a host of others applied scientific methods to police work. And *mirabile dictu*, they solved crimes.

Forensic detectives soon became a standard fixture in mystery literature. Jacques Futrelle's Professor S. F. X. Van Dusen, M.D., Ph.D., LL.D., F.R.S.; Arthur B. Reeve's chemistry professor Craig Kennedy; and others attempted to use scientific methods (to the extent that their authors understood them) to solve crimes. Even Dame Agatha Christie, though primarily a writer of "cozies," kept up on the latest information about rare poisons. The appeal of this new genre is spoken of by Hugo Gernsback in *The Writers' Digest* for February 1930:

> Scientific detection of crime offers writers the greatest opportunity and most fertile field since the detective first appeared in fiction. Radio, chemistry, physics, bacteriology, medicine, microscopy—every branch of science can be turned to account. The demand for this material is large, the supply is small.

As the editor of *Scientific Detective Monthly*, Gernsback surely knew whereof he spoke.

The popularity of these authors kept the public in mind of the possibilities of forensic science. This appetite was reinforced perhaps by the notoriety of several actual criminal cases of the period, some unsolved and some solved but in unsatisfactory ways. Among them are:

The 1874 kidnapping in Philadelphia of four-year-old Charlie Ross. The police, in a rarely equaled instance of misfeasance, malfeasance, and nonfeasance, squandered all opportunities to find the child or catch his kidnappers.

The 1888 "Jack the Ripper" mutilation murders of at least six prostitutes in London, unsolved because of police incompetence and lack of sound investigative techniques.

The 1892 Lizzie Borden case. In Fall River, Massachusetts, Lizzie was acquitted of the hatchet murders of her father and stepmother. Her guilt can be assumed—it is virtually impossible that anyone else had the required access to the crime scene. But sloppy police work in the early stages of the case prevents us from knowing for sure if Lizzie did it.

The 1920 armed robbery of a factory payroll in South Braintree, Massachusetts, for which the Italian anarchists Nicola Sacco and Bartolomeo Vanzetti were convicted and executed. Future Supreme Court justice Felix Frankfurter criticized the handling of the case, and there remains serious debate among crime experts as to the guilt of the two men.

The 1932 kidnapping of Charles Lindbergh's infant son, for which Bruno Richard Hauptmann was convicted and executed. There are still arguments in favor of his innocence. And, even if guilty, he certainly did not act alone. Yet no one else was ever apprehended or even accused.

This book delves into the history of criminalistics, a general term for the science of forensic investigation, which in turn is defined as the careful scientific examination of a crime scene to determine from the objects left behind just who did what, with what, and to whom. As one technical definition puts it, criminalists

> examine and identify physical evidence to reconstruct a crime scene. Physical evidence can be a weapon, a piece of clothing, a bloodstain, drugs, or even a vapor in the air. Criminalists use this physical evidence to provide a link between a suspect and the victim. The transfer of clothing fibers or hair fibers between a suspect and the victim can provide just such a link. Fingerprints, bullets, and shoe impressions are other important links. . . . Criminalists collect physical evidence at crime scenes and receive evidence at the laboratory, which has been collected at the crime scene by crime scene investigators. The proper collection

of evidence is essential to prevent contamination and destruction of the evidence. . . . Criminalists are often called to court to provide expert testimony regarding their methods and findings.

This distinguishes forensic science from criminology, the study of crime as a social phenomenon and the province of sociologists, psychologists, and penologists. Where criminologists theorize about the root causes of crime and the effectiveness of punishment versus rehabilitation, criminalists use scientific disciplines to analyze a crime scene and gather information for the purpose of apprehending and convicting a perpetrator, whatever his or her social or psychological background.

This book explores the history of the forensic sciences as well as the dedicated and sometimes obsessed people who developed the numerous forensic disciplines in use today. We'll look at how these methods came into being and why police investigators began applying them, however reluctantly, to solving crimes. Writing in 1908, the author Arthur Train, who had recently quit his job as district attorney for New York County, was a bit contemptuous of forensic detection:

> . . . No intelligent person to-day supposes that, outside of Sir
> Conan Doyle's interesting novels, detectives seek the baffling
> criminal by means of analyzing cigar butts, magnifying
> thumb marks or specializing in the various perfumes in favor
> among the fair sex, or by any of those complicated, brain
> fatiguing processes of ratiocination indulged in by our old
> friend, Mr. Sherlock Holmes. . . . The magnifying glass is not
> one of the ordinary tools of the professional sleuth, and if
> he carries a pistol at all it is because the police rules require
> it, while those cases may be numbered upon the fingers of
> two hands where his own hair and whiskers are not entirely
> sufficient for his purposes in the course of his professional
> career.

The past century has seen a great increase in the skills and capabilities of the forensic investigator, particularly after important breakthroughs in medicine and in the biological sciences. The criminalist today is one of a team of highly trained specialists. He or she uses the advances of many sciences and brings far more rapid and reliable solutions to cases than even two decades ago. Of course, this ideal is not always realized. Most crimes are still solved by what might be called good old-fashioned police work—asking questions, giving "deals for squeals," and leaning hard on the person with the strongest and most apparent motive. Forensic techniques are painstaking and time consuming. They require highly trained, dedicated technicians equipped with expensive tools. In the past, few police departments have had the personnel, the money, or the inclination to use their resources in this way. But with the federal government's new emphasis on homeland security, that situation is rapidly changing.

One of the revelations of modern forensic investigation is that good old-fashioned police work, despite its successes, produces an unacceptably high percentage of false convictions. This is one reason why in 2002 Governor George Ryan of Illinois, claiming that the state's death-penalty system was "fraught with errors," declared a moratorium on executions. In a series of highly publicized cases, DNA testing had demonstrated that thirteen of the prisoners awaiting execution on Illinois' death row were innocent of the crimes for which they had been convicted. Some of the exonerated prisoners had falsely confessed to their crimes, demonstrating that the pressure of a police interrogation does not always disclose the truth.

Forensic science is still rapidly evolving. Even the few years of the new millennium have seen major advances in DNA testing, blood-spatter analysis, the automated detection of explosives, and face-pattern recognition. On the other side of the coin, microscopic hair examination has been discredited; bite-mark evidence is, or should be, discarded; handwriting analysis has become suspect; toolmark comparisons are less reliable than previously thought; bullet composition comparisons have been shown to be bad science; and, after a hundred years of being the burglar's bane, latent fingerprint identification has been shown to be less foolproof than thousands of prosecutors and fingerprint analysts have sworn it to be. Problems are now coming to light with certain methods of arson investigation as well as other forensic specialties long thought to be reliable.

We will look at how these various errors—some of them the result of bad science—crept into courtrooms in the first place. We will also see how difficult it has been—and continues to be—to challenge them in court once they become entrenched. More frightening are the cases where the sworn testimony of supposed experts in hundreds of trials has proven to have more in common with witchcraft or voodoo than with responsible science.

Rigorously conducted experiments have demonstrated what criminal defense attorneys have long suspected—that eyewitness testimony (perhaps the most damning type of evidence in the eyes of the jury) is possibly the least reliable of all. Its accuracy rate is now thought to be less than 50 percent. We now know that when a witness points a quivering finger at a defendant and declares, "He's the one! I'll never forget that face—never!" he or she is mistaken more than half the time.

Recent Supreme Court rulings on the admissibility of expert testimony have placed a greater responsibility on judges to distinguish reliable from unreliable science. As different judges rule differently on what should and should not be admissible, this area of the law remains in a state of flux. It will take years and perhaps decades to bring order out of this chaos.

1 : All Our Yesterdays

> In future time, through all coming
> generations, let the king, who may be in the
> land, observe the words of righteousness
> which I have written on my monument; let
> him not alter the law of the land which I
> have given, the edicts which I have enacted;
> my monument let him not mar. If such a
> ruler have wisdom, and be able to keep his
> land in order, he shall observe the words
> which I have written in this inscription;
> the rule, statute, and law of the land which
> I have given; the decisions which I have
> made will this inscription show him; let him
> rule his subjects accordingly, speak justice
> to them, give right decisions, root out the
> miscreants and criminals from this land,
> and grant prosperity to his subjects.
> —*The Code of Hammurabi*, circa 1750 B.C.

FOR THERE to be a solution to a crime, forensic or otherwise, there must first be actions that are defined as criminal. The definition of crime varies from time to time, place to place, and context to context. Killing another human being is murder—except when it isn't, as in war or when mandated by the state as an execution.

Incest is a crime—unless you are the pharaoh of Egypt and are expected to marry your sister. For many centuries, harboring a runaway slave was a crime; today, throughout the world, slavery itself is illegal.

For most of history, a crime has been whatever the local ruler declared it to be. Local officials twisted the laws to suit themselves since few people knew just what was prohibited or just what punishments there were. Indeed the nobles of ancient Rome held on to their power partly by keeping others in ignorance of the laws they nevertheless enforced with great severity against the plebeian classes. After a few hundred years of this, even the nobles decided that the privilege was not worth continuing. An ambassador was sent to Athens to study Greek law and to draw up a definitive set of laws for Rome; and in 449 B.C. the Law of the Twelve Tables was published on twelve bronze tablets and set up in the Forum.

More than a thousand years earlier, Hammurabi, the king of Babylon from about 1796 to 1750 B.C. decided that everyone had a right to know what the laws were. This may have been a ploy to prevent his own nobles from making up the laws as they saw fit. In any event, he inscribed his body of laws, known today as the Code of Hammurabi, on seven-foot-high stone slabs that were then erected at various places around his kingdom. All of them have disappeared over the centuries, except for one that was found in 1901 by the Egyptologist Gustav Jéquier during an expedition to ancient Elam, a town in Khuzestan, a province of present-day Iran. On it Hammurabi begins by stating his authority:

. . . Anu the Sublime, King of the Anunaki, and Bel, the
lord of Heaven and earth . . . assigned to Marduk, the over-
ruling son of Ea, God . . . dominion over earthly man, and

made him great among the Igigi, they . . . called by name
me, Hammurabi, the exalted prince, who feared God, to
bring about the rule of righteousness in the land, to destroy
the wicked and the evil-doers; so that the strong should not
harm the weak; so that I should rule over the black-headed
people like Shamash, and enlighten the land, to further the
well-being of mankind. . . .

Then he got down to business with 282 specific injunctions,
some of interest to us:

> If any one ensnare another, putting a ban upon him, but
> he cannot prove it, then he that ensnared him shall be put to
> death [an apparent forerunner of "thou shall not bear false
> witness"].
>
> If any one bring an accusation against a man, and the
> accused go to the river and leap into the river, if he sink in
> the river his accuser shall take possession of his house. But if
> the river prove that the accused is not guilty, and he escape
> unhurt, then he who had brought the accusation shall be
> put to death, while he who leaped into the river shall take
> possession of the house that had belonged to his accuser
> [crime-solving by magic].
>
> If any one bring an accusation of any crime before the
> elders, and does not prove what he has charged, he shall, if
> it be a capital offense charged, be put to death. . . .
>
> If any one break a hole into a house [break in to steal],
> he shall be put to death before that hole and be buried. . . .
>
> If a man put out the eye of another man, his eye shall be
> put out [the now familiar "eye for an eye"].
>
> If he break another man's bone, his bone shall be broken.
>
> If he put out the eye of a freed man, or break the bone of
> a freed man, he shall pay one gold mina.

If he put out the eye of a man's slave, or break the bone of a man's slave, he shall pay one-half of its value.

If any one strike the body of a man higher in rank than he, he shall receive sixty blows with an ox-whip in public.

If a free-born man strike the body of another free-born man of equal rank, he shall pay one gold mina.

If a freed man strike the body of another freed man, he shall pay ten shekels in money.

If the slave of a freed man strike the body of a freed man, his ear shall be cut off. . . .

If a builder build a house for some one, and does not construct it properly, and the house which he built fall in and kill its owner, then that builder shall be put to death.

Sometime during the third century B.C., Hieron II, the tyrant of Syracuse, gave a quantity of gold to one of the city's goldsmiths. He was asked to fashion a gold laurel wreath as a crown for the winning athlete at an impending festival. For a reason long lost to history, Hieron came to suspect the goldsmith of secretly replacing some of the gold with an equal weight of silver—not a good trick to play on a tyrant. But was it true? Hieron asked the city's leading citizen, the polymath Archimedes, to devise a method to detect this possible debasing, preferably one that wouldn't involve the destruction of the wreath.

As told two centuries later by the Roman architect and historian Vitruvius, Archimedes saw the answer to his problem as he sank into his bathtub and noticed the water displaced by his body spilling over the edge. With a great cry of "Eureka!" he leaped from the bath and ran naked through the city streets.

Watching his body push the water from the tub, Archimedes had realized that an ounce of gold, being denser than an ounce of silver, would displace less water because it had less volume. If you picture an ounce of copper—eleven pennies—as opposed to an ounce of balsa wood—about the size of four decks of cards—you have an exaggerated image of the phenomenon. All Archimedes had to do was drop the crown in an amphora filled to the brim with water and then do the same with a lump of gold that weighed the same. If more water flowed out with the crown than the gold, it proved that something had been added to the mix.

Or he might place the crown on one side of a balance scale and enough gold onto the other pan to bring the scale into balance. The scale would then be immersed in water. If the crown were not pure gold, it would displace more water—the difference in buoyancy would cause the lump of gold to sink and the crown to rise. Whichever method Archimedes used, Vitruvius records that the crown was shown to be debased, and the goldsmith was beheaded. Archimedes might be the first European to use science to solve a crime.

Some fifteen hundred years later, in the fourteenth century A.D., we find the first recorded use of expert testimony in criminal trials in Europe. Among these earliest expert witnesses were "masters of grammar" who could read and interpret the medieval church Latin in which the laws were written and so determine the proper form for such swearing.

In a 1554 trial, *Buckley v. Rice*, the judge noted that "If matters arise in our law which concern other sciences or faculties, we commonly apply for the aid of that science or faculty which it concerns, which is an honorable and commendable thing in our law, for thereby it appears we do not despise all other sciences but our own, but we approve of them, and encourage them as things worthy of commendation."

Back then, the word "science," from *scientia*, the Latin word for knowledge, connoted "what there is to know," rather than the formal study of a particular field of knowledge. But the beginnings of forensics can be found in this four-centuries-old ruling.

THE WASHING AWAY OF WRONGS

> Among criminal matters, none is more serious than capital
> cases; in capital cases nothing is given more weight than the
> initially collected facts; as to these initially collected facts
> nothing is more crucial than the holding of inquests. In
> them is the power to grant life or to take it away, to redress
> grievances or to further iniquity.
>
> —Sung Tz'u, *The Washing Away of Wrongs,* 1247

As seems to be true of much of human knowledge, the roots of
scientific forensic investigation lie in ancient China. When Portu-
guese traders arrived in Canton in 1517 they were impressed by
how carefully and thoroughly Chinese judges examined the facts
of their criminal cases before reaching their verdicts, particularly
when compared to European practice of the time.

Robert van Gulik (1910–1967), a Dutch citizen who grew up
in Indochina and who spoke Mandarin Chinese, became interested in medieval Chinese detective stories when he was a member of the Dutch diplomatic mission in Peking. Once a popular
genre, they were almost unknown in the China of the 1940s. Van
Gulik began translating one of the classics, *Dee Goong An (The
Criminal Cases of Judge Dee),* into English in 1947 and completed it in 1949 while stationed in Tokyo. Set in T'ang dynasty
China, the tale is based on the career of Ti Jen-chieh, an actual
magistrate who lived from 630 to 700 A.D. and who was an as-
tute solver of complex crimes. The original stories about Judge
Dee and other famous magistrates on which van Gulik based his
tales were composed in the seventeenth century. Based on thou-
sand-year-old criminal cases, the tales show that the function of
the criminal investigator was well understood by the Chinese
long before it developed in the West. Judge Dee used a staff of
investigators when he didn't go forth in disguise himself.

The procedures of his time called for the examination of witnesses, suspects, and physical evidence. In order to determine the cause of death in possible murder cases, bodies were examined by specially designated coroners. Since a case could not be closed until a culprit confessed, a magistrate who was convinced of a suspect's guilt might judiciously apply torture to induce a confession. If the magistrate was mistaken and tortured the wrong man, he himself could be subjected to whatever torture he had inflicted on the suspect—a system that created a harsh but careful judiciary. In these seventeenth-century stories the magistrates also put great faith in the intercessions of ghosts and spirits to guide them to the truth.

Chinese forensic techniques and procedures were well established two thousand years before the original *Dee Goong An* was written. Archaeologists at a dig of a Ch'in dynasty tomb in Hubei Province in 1975 discovered a bundle of bamboo strips on which were inscribed a text that dates to the period known as the Warring States (475–221 B.C.). Compiled by then-chancellor Lü Pu-wei, the text, known as *The Spring and Autumn of Master Lü*, is a manual of forensic procedures. One section of it discusses how to examine a crime scene; one how to relate evidence found at the scene with the findings of the *lingshi* (the coroner); another how best to examine a corpse for broken bones and other trauma; and another how to determine the time of death from the condition of the body. In cases of death by hanging, the coroner recorded the sort of rope used, the structure from which it was hung, the position and location of the victim, and the state of the body. The victim's friends and relatives were questioned about the victim's affairs and asked to suggest the names of anyone who might have had a motive for the crime.

Establishing the exact cause of death, or the amount and type of trauma suffered by the victim if he lived, was vital—the type

and severity of punishment depended on the degree of harm inflicted. During the Sung dynasty (960–1279 A.D.), punishments
authorized by the state included beatings with the "light rod,"
beatings with the "heavy rod," imprisonment, and execution.
The Sung Code of 962 A.D. established the penalty for a severe
assault at forty blows of the light rod. But:

> If there are wounds or if weapons other than fists are used,
> the penalty is sixty blows of the heavy rod. If the wounds
> cause the loss of a square inch of hair or more, the penalty is
> eighty blows. If blood is drawn from the ear or eye, or there
> is the spitting of blood, one hundred blows.

A book called *Yi Yu Ji*, which apparently translates to *A Book
of Criminal Cases*, dating to the Wu dynasty (264–277 A.D.), relates that the coroner Zhang Ju investigated the case of a man
whose body had been burned in a fire. His wife was suspected
of killing him and of then setting the fire to cover her crime.
She denied it, of course, saying that the fire had been a horrible
accident.

Zhang Ju made a fire like the one that had killed (or not killed)
the man. In it he burned two pigs, one alive and the other dead.
When he examined the bodies of the pigs afterward, he found
that the live pig had ashes in its mouth while the already dead pig
did not. The victim had no ashes in his mouth and so, Zhang Ju
concluded, he had already been dead when the fire was set. The
woman confessed to murdering her husband.

In 995 A.D., Emperor T'ai Tsung decreed that in all cases of
suspected homicide or of death involving bodily injury, a coroner be appointed and an imperial inquest held. The coroner
was required to inspect the crime scene, conduct a postmortem
on the body, and report his findings to the imperial authorities. In his report he was required to include a front-and-back
diagram of the body with wounds or other markings indicated.

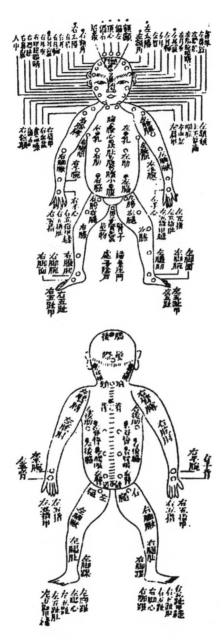

Under the Sung dynasty in China, coroners who inspected a crime scene were required to submit a front-and-back diagram of the body with wounds or other markings indicated.

This procedure formalized and standardized investigations into suspicious deaths.

Toward the end of the Sung dynasty, Sung Tz'u, a jurist who as a junior official had become known for successfully suppressing bandit gangs in several provinces, rose to the high office of judicial intendant. He wrote a book of criminal forensic procedures entitled *Hsi yüan chi lu* (*The Washing Away of Wrongs*). Published in 1247, it remained in print and in use for the next 650 years.

It includes the following advice:

A coroner must be serious, conscientious, and responsible.
He must personally examine each dead body or the
wounds of an injured person. The particulars of each
case must be recorded in the doctor's own handwriting.
A coroner must not avoid performing an autopsy because
he detests the stench of corpses. A coroner must refrain
from sitting comfortably behind a curtain of incense that
masks the stench, must not let his subordinates do the
autopsy unsupervised or allow a petty official to write his
autopsy report, leaving all the inaccuracies unchecked and
uncorrected.

Much in the writings of Sung Tz'u has a modern feel to it. In the case of a serious assault and battery, *The Washing Away of Wrongs* ruled that if the victim were seriously injured, a recovery period was to be set. If the victim died within this period, he was deemed to have died of his wounds, and the accused would then be tried for murder. If he lived beyond the recovery period or died of a cause that was obviously unconnected to the crime, only assault was charged. Then, ironically, the victim was placed in the care of the accused, the person with the strongest possible motive for wishing him a complete recovery.

Sung Tz'u did not have the highest opinion of the motives or behavior of his fellow men, nor even of his fellow bureaucrats. He suggested, for instance, that if the relatives of the victim did not wish to have an inquest, it was perhaps because they had been bought off by the accused. He believed that an attendant at an inquest who accepted a gift from anyone should be punished. He stressed that the officials conducting the inquest eschew personal relationships with anyone involved and avoid staying at the houses of the relatives of either the victim or the accused.

In some of Sung Tz'u's advice you can hear the exasperated sigh of the practitioner trying to drum some sense into the heads of his subordinates: "In writing up inquests, do not write 'The skin was broken. Then blood flowed out.' In general when the skin is broken, blood flows out."

His descriptions of what to look for in suspicious deaths are precise and detailed:

> When people have committed suicide by hanging, the eyes will be closed, the lips and mouth black, and the mouth open with the teeth showing. If hanged above the Adam's apple, the mouth will be closed, the teeth firmly set, and the tongue pressed against the teeth but not protruding. [This] can easily be distinguished from cases where the victim was strangled by someone else . . . with the death passed off as suicide. Where the victim has really killed himself by hanging . . . the flesh where the rope crosses over behind the ears will be deep purple in color. . . . If another man strangled the victim and tried to pass it off as suicide, the mouth and eyes will be open, the hands apart, and the hair in disorder . . . and the tongue will neither protrude nor will it be pressed against the teeth.

The appearance of the body after being beaten with various implements—fists, staves, whips, axes, bricks, and a few others—

is described. Detailed descriptions suggest what to look for in cases of drowning. Sung Tz'u differentiates between accidents, suicides, and murders. He describes the effects of various poisons. If the victim seems to have been in good health and there is no obvious cause of death, Sung Tz'u suggests that the official look for evidence of bamboo slivers inserted in the nose, ears, or under the fingernails, or for other objects forced into the mouth, rectum, or vagina. A careful search is to be made of the scalp to make sure that a nail hasn't been driven through the skull and into the brain.

Sung Tz'u records the case of a particularly brutal murder in which a peasant was hacked to death with a rice-cutting sickle, a weapon that suggested an assault by a fellow peasant. The magistrate gathered the peasantry in the village square, where he inspected their sickles. There was no sign of blood on any of them. So he had them place their sickles on the ground. After a while the blowflies—shiny green flies with small spots of orange on their thoraxes—began to fly in random patterns around the sickles. These flight paths became increasingly less random, until shortly, most of the flies landed on one particular sickle. As carrion-lovers, the flies were attracted to bits of flesh and blood too small for the human eye to detect. The guilty peasant was led away, and the first recorded case of forensic entomology entered the record book.

The Washing Away of Wrongs remained the high point of forensic investigative techniques until modern scientific methods superseded classical Chinese empiricism. Yet the ethical standards demanded by Sung Tz'u are no less important today. Perhaps his words should be engraved over the entrance to every forensic science lab: *A coroner must be serious, conscientious, and responsible.*

2 : Beginnings

FOR MUCH OF European history, from the dissolution of the Roman Empire to the seventeenth and eighteenth centuries, crime-solving presented no great difficulty. One merely accused someone of the offense and then tortured him until he confessed. Actual guilt or innocence was not allowed to pervert the process. As Enrico Ferri explained:

> The tortures, which we incorrectly ascribe to the mental brutality of the judges of those times, were but a logical consequence of the contemporaneous theories. It was felt that in order to condemn a man, one must have the certainty of his guilt, and it was said that the best means of obtaining this certainty, the queen of proofs, was the confession of the criminal. And if the criminal denied his guilt, it was necessary to have recourse to torture in order to force him to a confession which he withheld from fear of the penalty. The torture soothed, so to say, the conscience of the judge, who was free to condemn as soon as he had obtained a confession.

But how to determine whom to torture? Hans Gross, in his landmark 1906 book *Criminal Investigation*, tells of a medieval practice that was still in use in the German countryside when he was an investigator. Called the "hereditary sieve," it was used to reveal the identity of a thief. Beans (stolen if possible) were thrown into a sieve one by one. A name was pronounced with the toss of each bean. If the bean jumped out, the named man was innocent; if the bean stayed in, you had caught your thief. There were alternate methods of handling the sieve, all of them equally effective.

Sometimes measures were used that would be equally frowned upon today. In the 1830s in the county of Pest, in Hungary, Count Gideon Raday was called upon to stop a rash of robberies that were causing great distress among the people. Not having any modern forensic techniques at his disposal, he merely hanged the mayor of the town that lay at the center of the epidemic. The robberies ceased.

Arden of Feversham, first published in 1592, is the story of Thomas Arden (or Arderne), a successful burgher of the English town of Feversham (today Faversham), and of his wife Alice, who repeatedly tried to murder him and finally succeeded. Although no writer is credited for the play, Shakespeare may have written it. The poet Swinburne thought so. And he may have even played one of the two hired murderers, the one called Shakebag. Others hold out for Marlowe because that play is a tough-minded story with no hero—the sort of thing Marlowe enjoyed. The strongest claim for authorship, based on an analysis of word usage, goes to Thomas Kyd, who along with Shakespeare and Marlowe was one of the leading dramatists of the day.

The tale is taken from Holinshed's *Chronicles* (1577), which tells that one Thomas Arden was indeed murdered on Sunday, February 15, 1551, at his home in Feversham, Kent. The story may not have a hero, but it has a detective—and, for the time, a first-class one at that.

Thomas Arden, "a man of a tall and comely personage," married Alice Brigandine, "a gentlewoman who was young, well shaped, and every way handsome." He then proceeded to ignore

her and devote his energies to acquiring properties recently ex-
propriated by Henry VIII from the Catholic church. Alice soon
directed her energies elsewhere, finding solace in the arms of a
tailor named Thomas Mosby, who visited her often, and as the
Chronicle would have it, "lay with her and . . . kept her in abus-
ing her body."

Arden seemed to have little objection to this. He invited
Mosby to stay over while he went hither and yon in search of
church properties he could add to his collection. To those of us
who seek illicit romance, this might seem an ideal situation. But
it was not so to Mistress Alice and her lover. They determined to
do away with the often-absent Master Arden. The first attempt,
as far as we know, was with a poison that Alice purchased from a
painter in Feversham. He told her to put it in a bowl of porridge
and pour milk over it, but she put the milk in first and poured the
poison over that. Arden thought the stuff tasted foul after having only a spoonful or two of it, and Alice tossed it out before he
could ask any questions. Still, a couple of hours later Arden "fell
into extreme purging, upward and downward." He survived the
experience.

After several more bungled attempts, the lovers hired a pair of
brigands—Black Will and Loosebag (the Shakebag of the play)—
to do the deed for ten pounds.

One Sunday evening when Arden was in the parlor with
Mosby and the two were hunched over a game of tables, a precursor to backgammon, Black Will concealed himself in a corner.
As recounted in the play, Mosby then punned, "Now may I take
you, sir, if I will," and events unfolded as follows:

> "Take me?" quoth Master Arden. "Which way?"
> With that Black Will stepped forth and cast a towel
> about [Arden's] neck so as to stop his breath and strangle
> him. Then Mosby, having at his girdle a pressing iron of

fourteen pounds' weight, struck him on the head so that he fell down and gave a great groan. Then they carried him into the counting house, where—the pangs of death coming on him—he gave another great groan and stretched himself. Black Will gave him a gash in the face and took the money out of his purse and the rings from his fingers. "Now this feat is done," he said, as he came out of the counting house; "give me my money." Mistress Arden gave him ten pounds, and . . . he rode away.

Alice Arden then stabbed the body of her dead husband a few times before returning to the dining room and calmly sitting down to dinner with two London grocers curiously named Prune and Cole. She told them not to wait for her husband as he might be delayed. When the visitors returned to their rooms in the local inn (the aptly named Flower-de-Lice), Mistress Arden sent her servants out "to inquire for her husband in divers places." In their absence she, along with her own daughter and Mosby's sister, dragged the body out through the new-fallen snow and left it in a nearby field.

The clever Mistress Arden then set up such a-weeping and a-wailing about her absent husband that the mayor of Feversham was pulled out of bed. He formed a group to hunt for the missing man. As it happened, Prune the grocer spotted a man-shaped lump in the field outside Arden's house, saying, "Stay, for methinks I see one there."

Then the mayor showed qualities that would have done Sherlock Holmes proud. He prevented everyone from approaching the body and examined the footprints in the snow. Three pairs of footprints led from Arden's garden door to the place where he lay, and three pairs of footprints led back to the door—the only footprints since the snow had fallen. Three people had left the house with Arden's body, the mayor concluded. They had dropped it where it then lay, and returned to the house. The mayor's forensic genius lay in his having the wit to prevent his men from trampling the area, and in examining it closely, probably in the dim light of oil lamps, to determine the facts.

The mayor had the house searched. Bloody rags were found in the parlor and a bloody knife was discovered in the bedroom. Within two hours of committing their well-planned murder, Alice Arden, her daughter, and her maid were arrested and taken to jail.

Then the mayor and his men went to the Flower-de-Lice and awoke Mosby. When they found blood on his stockings and purse, he joined his paramour in jail. All were tried at the next assizes in Feversham and variously hanged or burned to death along with a man named Green, who had had the bad luck to be mentioned in a letter from Mosby to Alice. In fact he had no knowledge of the crime and had taken no part in it. His innocence was established several years later.

In 1828, Eugène François Vidocq, newly retired as the head of the Sûreté, the detective branch of the Paris police, wrote his memoirs, offering the European public its first view of the life of a police officer. Vidocq had spent the first half of his life as a criminal and therefore understood criminals well. As a detective, he made most of his arrests by wandering in disguise among the criminal classes in Paris and listening to their conversations. His knowledge of criminal methods helped him anticipate, and thus foil, many crimes. As he put it:

> Each day increased the number of my discoveries. Of the many who were committed to prison, there were none who did not owe their arrest to me, and yet not one of them for a moment suspected my share in the business. I managed so well, that neither within nor without its walls, had the slightest suspicion transpired. The thieves of my acquaintance looked upon me as their best friend and true

comrade, the others esteemed themselves happy to have
an opportunity of initiating me in their secrets, whether
from the pleasure of conversing with me, or in the hope of
benefiting by my counsels.

Vidocq is credited with introducing the first card-indexing
system into police procedures, and of being the first to make
plaster of paris shoe and foot impressions. In addition to publishing his autobiography after leaving the police force, he set up
a paper-manufacturing plant and patented an unalterable bond
paper and an indelible ink. In 1833 he founded *Le bureau des
renseignements* (The Office of Intelligence), the world's first private detective agency.

Vidocq is thought to be the inspiration for Edgar Allan Poe's
C. Auguste Dupin, the world's first fictional detective, as well
as Émile Gaboriau's detective, Monsieur Lecoq. Today the Vidocq Society, an exclusive nonprofit, crime-solving association
of forensic experts, meets in Philadelphia and takes up unsolved
criminal cases of interest to its members.

The prosecution that resulted in the earliest known wrongful
conviction in a murder case in the United States suffered from
several of the problems that still plague us today—inflamed local
sentiment, flawed "expert" testimony, and reliance on the evidence of a "jailhouse snitch."

In 1812, Russell Colvin, who lived and worked on the Manchester, Vermont, farm of his father-in-law, Barney Boorn, disappeared. Seven years later, his wife's uncle Amos had a vision.
Russell appeared to Amos in a dream and said that he had been
murdered and his body dumped into a cellar in a potato field on
the farm. The hole was dug up, but Russell's ghost had apparently
been mistaken—his body was not there. There were broken dishes,
two knives, and a button, but no Russell. But Russell's wife, Sally,

claimed to recognize the contents as having been her husband's. Sally had a strong motive to prove Russell dead—she had given birth a few years after Russell's disappearance. As long as Russell was presumed to be alive, even if long absent, he was the child's presumed father. If Sally were ever to receive support from the child's actual father, her husband would need to be dead.

A short time later the Boorns' sheep barn burned to the ground. Not long after that, a dog dug up a couple of partial bones, whereupon three separate doctors promptly identified them as human.

Popular suspicion settled on Sally's brothers, Jesse and Stephen. They were believed to be Russell's murderers, the despoilers of the potato cellar, and the burners of the barn. The brothers were known to have disliked Russell, whom they looked upon as a wastrel who spent too much time drinking at the local tavern and too little time working in the fields. The popular logic worked backward from the discovery of the bones: Why were the bones buried there? Because they'd been moved out of the barn for some unknown reason. Why had the barn been burned? To hide any trace of the transfer of the body. Why had it been in the barn in the first place? It was moved there from the potato cellar. And, of course, it had been placed in the potato cellar in order to hide the murder.

On the basis of this retrograde logic, Jesse was promptly locked up and an arrest warrant issued for Stephen, who had since moved to New York. Jesse was placed in a cell with a forger and fink named Silas Merrill. He promptly went to the authorities to find out what they wanted to hear, then promptly told it to them with an added flourish. Jesse had confessed all to him, he said: the brothers had argued with Russell, Stephen had hit him with a club, and their father, Barney, who happened along at that moment, had borrowed a pen knife from Stephen and cut Russell's throat. Then the three of them buried Russell in the potato cellar;

later they dug him up and reburied him in the barn. After the barn burned, they moved the body to the location where the dog had found the bones. At least in this version they hadn't deliberately burned the barn. Murderers, yes—but not barn burners.

Silas Merrill's price for his testimony against the Boorn brothers was his immediate release. State's Attorney Calvin Sheldon was willing to trade a forger for a trio of murderers any day.

Jesse then confessed, declaring that Stephen had committed the actual murder, and that his father had had nothing to do with it. What pressures were put on him or what promises made can only be guessed. But when Stephen—to everyone's surprise—returned voluntarily from New York and protested his innocence, Jesse recanted his confession.

But the evidence piled up. Seven years after the event the Boorns' neighbors then remembered threats the brothers had made against Russell before the man went missing. They recalled sly comments the brothers made afterward, suggesting that they knew more than they would say about his disappearance. Two men suddenly remembered that they had seen Stephen and Russell fighting on the day Russell disappeared, though neither had stayed to see how the fight turned out.

With all the evidence piling up against him, Stephen now confessed in order to spare himself the death penalty, claiming that he had killed Russell in self-defense. The confession seems to have been written by Stephen's lawyers, since it used words and phrases certainly beyond his intelligence and level of education. Stephen was known to be rather slow.

So we have the ghost of Russell Colvin appearing to Uncle Amos, his bones dug up by the dog, the confessions of each of the two brothers, the story of the jailhouse snitch, and seven-year-old memories of various local people. Certainly enough to convict.

By the start of the trial, the evidence of the bones had disappeared—on closer examination, the physicians concluded that they were animal bones. In spite of the fact that there was no body (not even a bone), that the confessions had been recanted, and that the only evidence of substance was the story that one of the brothers had been seen fighting with the missing man, the brothers were convicted and sentenced to hang. Jesse's sentence was subsequently changed to life imprisonment. Colvin's ghostly appearance to Uncle Amos was inadmissible as evidence, but it was surely on the juror's minds.

Then serendipity took a hand. An article about the murder in a New York paper in November 1819 recounted how the ghostly appearance of Russell Colvin to Uncle Amos had set events in motion that led to the trial and conviction of the brothers. A New Jersey man, Tabor Chadwick, was in the lobby of a New York City hotel when someone near him began reading the article aloud. He immediately sat down and wrote one letter to the newspaper and another to the postmaster in Manchester, Vermont. It happened that he knew a man named Russell Colvin who worked as a farmhand in Dover, New Jersey. And this Russell Colvin talked a lot about Vermont, where he had come from.

It took a little coaxing to get Colvin to return to Vermont, but he did and was immediately recognized by his former neighbors. When the brothers were released from prison, Stephen was within six weeks of his execution date.

On October 7, 1925, Henry Sweet, twenty-one, and Carmen Wagner, his seventeen-year-old girlfriend, left their homes in Eureka, California, to go deer hunting in Coyote Flat, an area some forty-five miles to the southeast. Four days later the body of Henry Sweet was found in a deserted cabin. Twelve days later, on October 23, the girl's body was found at Baker Creek, a few miles away. She had been shot twice, and an attempt had been made to bury her body in a shallow grave. Dried blood and skin scrapings were found under her fingernails.

The police decided to arrest a couple of "half-breeds," Jack Ryan and his half-brother, Walter David, for the murders. David was picked up on the October 23 and Ryan the next day. There was absolutely nothing to connect them to either the killings or the victims, except that they lived in the area. David was released a couple of days later, but Ryan was held for trial in the murder of Wagner.

Ryan's trial began in February 1926 and lasted five weeks. In the end, the all-male, all-white jury found him innocent. But the citizens of Humboldt County were unhappy with the verdict. In January 1927 a local attorney and bootlegger, Stephen Earl Metzler, ran for the office of district attorney on a platform of righting this obvious injustice.

When Metzler was elected he set about to keep his promise to the voters. Instead of "Justice for all," his motto seems to have been "Give the people what they want."

On October 31, Walter David was found strangled to death with barbed wire. His body showed signs of torture. No one was ever tried for his murder. Shortly after the murder, Jack Ryan began to receive anonymous letters threatening him with a similar fate unless he confessed to the killings of Sweet and Wagner. The letters were being sent by District Attorney Metzler.

On July 12, 1928, Ryan was arrested and accused of the statutory rape of a thirteen-year-old girl. To avoid a trial he immediately pled guilty to two of the three counts against him. Held overnight in the local jail and interrogated by Metzler, by morning he had confessed to the two murders. Since he couldn't be tried a second time for the killing of Carmen Wagner, he was charged with the murder of Henry Sweet. He pled guilty, was sentenced to life in prison without a trial, and was shipped off to San Quentin that same day.

In 1947 the Bureau of Indian Affairs began an investigation into Jack Ryan's case. Stephen Earl Metzler admitted to the agents that he had set Ryan up for the rape charge, paying the girl's mother $100 for her false testimony. He also admitted that the murder confession had been beaten out of Ryan, and that the man who was probably guilty of the murders, Bill Shields, had actually provided him with information to set Ryan up. Still, it took six years to win Ryan a parole—not a pardon—from the State of California. On May 11, 1953, after twenty-five years, Jack Ryan walked out of San Quentin, not quite a free man.

On March 20, 1969, Governor Ronald Reagan commuted Ryan's life sentence to time served, effectively releasing him from parole. But he was still legally guilty. Ryan died in 1978.

It wasn't until 1996, after the culmination of an extensive unofficial investigation conducted in his spare time by Humboldt County detective Richard H. Walton, that Ryan's innocence was officially accepted. California governor Pete Wilson issued the first-ever posthumous pardon and exoneration, saying it was clear that Ryan had been framed by Metzler. In his declaration Wilson said:

> Unfortunately, we cannot do justice for Jack Ryan, the man. But we can do justice for Jack Ryan, the memory. And by doing so, we breathe vitality into our system of justice. We must remember that a just society may not always achieve justice, but it must constantly strive for justice. This means that we must not excuse the guilty nor fail to exonerate the guiltless. Therefore, so that justice is maintained, I grant Jack Ryan posthumously a pardon based on innocence.

3 : Who Was That Masked Man?

> If nature had only one fixed standard for the
> proportions of the various parts, then the
> faces of all men would resemble each other
> to such a degree that it would be impossible
> to distinguish one from another; but she has
> varied the five parts of the face in such a
> way that although she has made an almost
> universal standard as to their size, she has
> not observed it in the various conditions to
> such a degree as to prevent one from being
> clearly distinguished from another.
> —Leonardo da Vinci

LET'S LOOK AT some of the problems raised by the seemingly
simple problem of telling one person from another. In *The Art
of Cookery*, an eighteenth-century cookbook by Hannah Glasse,
the recipe for her allegedly delicious Jugged Hare begins, "Take
your hare when it is cased," meaning "First catch your hare." The
same problem exists with felons: they cannot be jugged until they
are caught, and they cannot be caught until they are identified.

From the Middle Ages to modern times, felons who were
not summarily executed (and there were hundreds of crimes

for which the penalty was death) were branded. The purpose of branding was for identification—to warn good citizens to beware the offender—rather than for punishment. It was thought to be insufficiently painful in an era when the rack, the thumbscrew, and the whip were in common use. In France, from the four-teenth century on, the *fleur de lis* was branded on the shoulder of a released convict. In eighteenth-century Britain, thieves who escaped hanging were branded on the cheek. In tsarist Russia, prisoners sent to Siberia were branded on both cheeks and the forehead. The practice did not die out around the world until the early twentieth century. It endured in China until 1905.

In various parts of the world, another form of identification by disfigurement was in use until quite recently—mutilation, usually ear or nose cropping, and castration. For many centuries China practiced amputation of the nose or feet. The practice of cutting off a thief's hand, still the law in Saudi Arabia and other Arab states, serves the triple function of identification, punish-ment, and deterrence.

Branding and mutilation were generic solutions, however. They were not very effective in identifying specific individuals. One would know that a person sporting a brand, a neatly re-moved nose, or a V-shaped cut in his left earlobe was a felon, but not know just which felon he was. A positive means of identifying a specific criminal was needed, and it would be many centuries before it was found. In the Roman Empire, written descriptions of missing criminals and runaway slaves were distributed. These focused on many of the same details found in the *portrait parlé*, a method developed in the nineteenth century by Alphonse Bertil-lon and still used by some police forces today. But these were no more accurate than the powers of observation and description of the writer.

Eugène Vidocq, the reformed felon who became head of the
Paris police in the 1820s, realized the importance of personal
identification. In his memoir he wrote:

> I was no sooner the principal agent of the police of safety,
> than, most jealous of the proper fulfillment of the duty
> confided to me, I devoted myself seriously to acquire the
> necessary information. It seemed to me an excellent method
> to class, as accurately as possible, the descriptions of all
> the individuals at whom the finger of justice was pointed.
> I could thereby more readily recognize them if they should
> escape, and at the expiration of the sentence it became more
> easy for me to have that surveillance over them that was
> required of me. I then solicited from M. Henry authority
> to go to Bicetre with my auxiliaries, that I might examine,
> during the operation of fettering, both the convicts of Paris
> and those from the provinces, who generally assemble on the
> same chain.

With the growth of cities in the nineteenth century and the
establishment of professional police forces came attempts to sys-
tematize the identification process. As penitentiaries became the
preferred places of rehabilitation, the idea emerged that repeat
offenders were insufficiently penitent and should receive harsher
sentences. Professional criminals facing these harsher sentences
would then naturally go to extreme lengths not to be identified.
Policemen, particularly detectives, were encouraged to attend
weekly criminal parades in which all the suspects in custody were
lined up. They would then stare at the faces of those passing
through the system so that they might recognize them when they
encountered them again. Visiting policemen from other jurisdic-
tions were also expected to attend these local lineups so that they
could memorize the features of local criminals and spot any felons wanted on outstanding warrants in their own cities.

It was the custom in New York City for detectives attending the lineups to wear masks so that the criminals could not return the compliment by recognizing *them.*

On occasion these efforts to identify wanted criminals and repeat offenders had unanticipated results. In July 1844 the prefect of the Paris police offered a reward of twenty francs for the identification of "recidivists," as they were called. Often a felon who was certain that he would be recognized struck a deal with a friendly policeman who would then turn him in and split the reward.

But personal identification for police purposes has serious flaws. Many freshman psychology courses demonstrate the dangers of eyewitness testimony by having someone run unexpectedly into a classroom, fire a revolver, then run out, or perform some other attention-grabbing stunt. The professor then asks the students to describe the actions and the actor. Seldom does anyone get all, or even most, of the salient facts correct.

It is also true that people may not notice something that is right in front of them. On the internet you can find a wonderful four-minute clip of a basketball-tossing game. As an exercise in observation, you are asked to count the number of times the ball is passed between the players wearing white shirts. After coming up with the number, you are asked if you noticed anything else. So you play the clip again, this time on the alert for something even slightly strange. And this time you see what initially escaped your view—a woman (the website explains) in a gorilla suit entering from the right, walking between the players, thumping her chest at the camera, and exiting on the left.

Not only in the classroom is eyewitness testimony problematic. In 1803 a New York City carpenter named Thomas Hoag, happily married and the father of a young daughter, suddenly disappeared. Two years later his sister-in-law heard his distinctive lisping voice behind her on the street. Turning, she saw that the man behind her was indeed Hoag. She pointed him out to the authorities, who then arrested him for deserting his family. At his trial, eight people, including his landlord, his employer, and a close friend, identified him. He had a familiar scar on his forehead and a recognizable wen on the back of his neck.

But the defendant insisted he was in fact one Joseph Parker. To prove it, he brought in eight witnesses of his own, including a wife of eight years. The judge could not decide.

Hoag's friend, with whom he had once exercised daily, remembered that Hoag had a large scar on the bottom of his foot. The defendant was asked to remove his boots, which he gladly did. There was no scar, and so Parker went home to his wife. Hoag was never found.

In Great Britain in 1877 there was a more serious case of mistaken identity. A man calling himself "Lord Willoughby" went to prison for defrauding persons whom the authorities referred to as "women of loose character." Since the man was clearly not entitled to the name he claimed, the prison records settled on the name "John Smith" as a suitable identifier.

In 1894, less than a year after his release from prison, a cluster of women, "mostly of loose character," complained to the police that they had been defrauded by a man who called himself "Lord Wilton," or, on occasion, "Lord Winton de Willoughby." Their descriptions of the man varied, but the bad checks the women had received all seemed to have been written in the same hand. About a year later, in December 1895, one of the ladies, Ottilie Maissonier, passed a Norwegian mining engineer named Adolf Beck on Victoria Street in London. She recognized Beck as Lord Winton de Willoughby and reported him to a policeman.

Beck protested his innocence, but the bobby took him back to the police station, where several other similarly defrauded women came to look at him. They too identified Beck as the bogus lord. A retired police constable who had dealt with the earlier "John

Smith" was called in to look at Beck, and he also swore that Beck and Smith were one and the same man. His opinion was confirmed by a second officer.

Beck was convicted and sentenced to seven years in prison, a harsher sentence than he would otherwise have received had he not been listed in the criminal records as a repeat offender. In 1896, Beck's lawyer managed to have the case reexamined on the grounds that the prison's own records showed that John Smith had been circumcised, whereas Beck had not. The Home Office decided not to grant him a new trial but did order his previous conviction expunged from the prison record.

After he had served his term and had been a free man for almost three years, the unfortunate Beck was arrested once again on new complaints that read a lot like the old ones. This time he was convicted and sentenced as a repeat offender. But just as he was about to be sent away to prison, a man calling himself Thomas was charged with offenses much like those for which Beck had just been convicted. When confronted with Thomas, the women who had identified Beck realized that they had made a mistake.

The police, now suspecting the truth, brought in witnesses to both earlier crimes. Thomas was identified as the real "Smith" and as the man who had committed all the crimes of which Beck had been convicted. Beck was at once granted a "free pardon" and awarded £5,000 compensation.

Eyewitness identification is not helped by the passage of time. In 1981, John Demjanjuk, a sixty-two-year-old Ukrainian who had emigrated to Cleveland, Ohio, in 1952 and worked as a steelworker, lost his United States citizenship for allegedly entering the country under false pretenses. Then, in 1986, he was extradited to Israel to stand trial for major war crimes. He was accused by the Israelis of being "Ivan the Terrible," a guard at the Treblinka death camp who had supervised the gas chamber and was responsible for the deaths of tens of thousands of Jews. In 1988, after the emotional testimony of Treblinka survivors who identified Demjanjuk as the monstrous guard, and despite his vehement denials, Demjanjuk was convicted and sentenced to death—a punishment that in Israel is reserved for those who commit crimes against humanity.

In 1993 the Israeli Supreme Court reversed the conviction. New evidence from Soviet archives showed that while Demjanjuk was probably a guard at Sobibor, another death camp in Poland (a charge he also vehemently denies), he was never in fact at Treblinka and was not Ivan the Terrible. The survivors' identifications of Demjanjuk as the guard they had seen under stressful conditions forty years earlier, though made in good faith, were mistaken. Demjanjuk may indeed have been guilty of war crimes, but not of those charged against him. In 2009 he was deported to Germany, where officials are considering trying him for yet other war crimes.

An infrequent but interesting and complex problem of identity concerns who a person is not rather than who he is. The most famous case of this kind during the last century was that of the missing Grand Duchess Anastasia of Russia and of the several women who claimed to be her.

Tsar Nicholas II, the last of the Romanov dynasty, was murdered along with his entire family on July 17, 1918, by the new Bolshevik government. Their bodies were burned and buried in a pit, the location of which was kept secret. But a persistent rumor had it that Anastasia, the youngest of the tsar's children, had escaped the massacre and was living under an assumed name in a foreign country.

Of the several claimants to her identity, the most convincing was a girl who jumped off a bridge in Berlin on February 17, 1920, one year and seven months after the mass execution. She claimed to have lost her memory, and she subsequently spent two years in a mental hospital where all the while people remarked on how closely she resembled the beautiful missing Anastasia. For want of a better name she called herself Anna Anderson. She claimed to have vague memories of another life, a grand and wonderful life cut short by tragedy.

She remembered being bayoneted, and being rescued by a soldier named Tschaikovsky who took her to Romania. They were married and had a child. When Tschaikovsky was killed in a street fight, she sent the child to an orphanage and gathered her courage to go to Berlin and ask Anastasia's aunt, Princess Irene, for help. It was then that she lost hope and jumped off the bridge.

For the rest of her long life (she died in 1984), she attracted both adherents and detractors from among Anastasia's royal relatives and the scientists and experts who examined her. Crown Princess Cecilie, the kaiser's daughter-in-law and a distant relative of Anastasia, believed in Anna, but the princess's son, Prince Louis Ferdinand, did not. Pierre Gilliard, who had been Anastasia's tutor, believed at first and then later changed his mind. Grand Duke Alexander, cousin of the tsar, firmly believed.

Grand Duke Ernst of Hesse, who did not believe, conducted an investigation which concluded that Anna was actually Franziska Schanzkowska, a Polish factory worker who had disappeared in 1920. But Anna claimed to know of a secret trip that the grand duke had made to Russia in 1916 to visit the tsar. Ernst adamantly denied making any such trip, but in 1966 the kaiser's stepson swore in court that he had been told that Ernst had indeed made such a trip. If Anna was an imposter, how could she possibly have known this?

The burial place of the Russian royal family was discovered and the remains removed in 1991. DNA testing confirmed that the grave had contained the bones of the tsar, his wife, and three of the children. Anastasia's remains were not found. At yet another site were found charred bones that may be those of Anastasia and her younger brother.

> Some men has plenty money and no brains, and some men has plenty brains and no money. Surely men with plenty money and no brains were made for men with plenty brains and no money.
>
> —From the notebook of Arthur Orton

In the days before fingerprinting, photography, and Bertillonage, simple identification could be a confounding exercise. When a mother says that she recognizes her son, no matter how unlikely the body in which she finds him, how are we to argue?

The cargo schooner *Bella* was one of the first vessels to disappear in what has since become known as the Bermuda Triangle. It set sail from Rio de Janeiro, bound for New York City by way of Kingston, Jamaica, on April 20, 1854, carrying a cargo of coffee beans, a crew of forty, and one passenger—twenty-five-year-old Roger Charles Doughty Tichborne, heir to one of the oldest and richest estates in England. But the ship never arrived at either port, and neither did Mr. Tichborne or any of its crew. A bit of wreckage was found, but no further trace of the ship. Lloyd's wrote it off as "foundered with all hands."

In July 1855, Roger Tichborne was formally declared lost at sea. It should have ended there, but Roger's mother, Lady Tichborne, was not one to give up easily. She placed ads in newspapers throughout Britain, America, Europe, and Australia in which she asked for word of her son. She described him as being rather undersized and delicate, with sharp features, dark eyes, and straight black hair. Many young men answered these ads, but none of them proved to be the missing Roger.

In 1862, when Roger's father died and his baronetcy passed to Roger's younger brother, Lady Tichborne began a fresh spate of advertising, offering a reward for information. An ad posted in the Melbourne, Australia, *Argus* read:

> A handsome reward will be given to any person who can furnish such information as will discover the fate of Roger Charles Tichborne. He sailed from Rio Janeiro on the 20th of April 1854 in the ship *La Bella*, and has never been heard of since, but a report reached England to the effect that a portion of the crew and passengers of a vessel of that name was picked up by a vessel bound to Australia, Melbourne it is believed. It is not known whether the said Roger Charles Tichborne was among the drowned or saved. He would at the present time be about thirty-two years of age, is of a delicate constitution, rather tall, with very light brown hair, and blue eyes. Mr. Tichborne is the son of Sir James Tichborne, now deceased, and is heir to all his estates.

This appeal of course occasioned a new gush of sightings, claims, and stories. In 1865, Lady Tichborne was notified by the Missing Friends Bureau of Sydney, Australia, that a man answering her description had been found in the town of Wagga Wagga in New South Wales, where he ran a butcher shop and answered to the name of Thomas Castro. He said, however, that this was not the name given him at birth. A second letter confirmed that Castro was in fact a British nobleman in disguise, and that he had admitted to at least one person that he was really Roger Tichborne. Mama sent Andrew Bogle, an old family retainer, and Michael Guilfoyle, the erstwhile head gardener at Tichborne Park, both of whom had moved to Australia, to interview the butcher. Castro in the meantime had done his best to learn everything he could about Tichborne and the family history. Somehow, both Bogle and Guilfoyle accepted this fat, barely literate man as the slim, elegant, well-read Roger Tichborne.

The newly minted young Tichborne then wrote to his "Dear Mama" directly, proving himself unable to write a grammatical English sentence and to have trouble in spelling certain words, among them "Tichborne" (he spelled it "Titchborne"). He assured his mama that he had the birthmark (of which she had no recollection), and recalled an incident at Brighton (which she did not remember).

But poor Roger had been brought up in France, after all, and he always did have trouble with his English. And he never could spell very well. Almost convinced because she desperately wanted to be convinced, Mama sent him a return ticket to England. As she wrote to her Australian contact, "I think my poor, dear Roger confuses everything in his head, just as in a dream, and I believe him to be my son, though his statements differ from mine."

But the man who returned was far from the son who had left. As Edward H. Smith put it, in *Mysteries of the Missing*:

> She had sent away, thirteen years before, a slight, delicate, poetic aristocrat, whose chief characteristic was an excessive refinement that made him quite unfit for the common stresses of life. In his stead there came back a short, gross, enormously fat plebeian, with the lingual faults and vocal solecisms of the cockney. In the place of the young man who knew his French and did not know his English, here was a fellow who could speak not a word of the Gallic tongue and used his English abominably.

Of course she recognized him as her son immediately. They met in Paris, thus avoiding the young man's British relatives until he had an opportunity to learn what he had to learn. He came, stayed, and learned fast. And Lady Tichborne was happy with her newfound son until she died some three years later in 1868.

The turns and twists of this story have filled at least a dozen books. Here we will follow the main thread until it reaches a sort of conclusion. In 1870 the Tichborne pretender filed suit to retrieve "his" birthright: the Tichborne estates and the baronetcy that went with them. He had used the intervening years to study up on everything Tichborne and to make friends with all the people whom he had supposedly known from childhood, using each new bit of information he unearthed as a lever to pry loose the next bit. It was noted that his handwriting grew more and more like that of the Roger Tichborne who had been on the *Bella*. Was he becoming more like his old self, or had he merely been practicing?

With the death of Lady Tichborne, the claimant had been cut off from his money supply, and the present heirs would not support his efforts to take the estate away. The case attracted a great deal of public attention, with the man in the street siding with the claimant. As the public saw it, greedy relatives were attempting to hang on to money that wasn't rightfully theirs. After all, a mother should know her own son, shouldn't she?

On May 11, 1871, a trial began that continued until March 1872. Sir John Coleridge, representing the family, questioned the chubby claimant for twenty-two days. More than a hundred witnesses took a hundred days to tell the jury that they knew Roger Tichborne, and that "that man over there" was he. Among the witnesses for the claimant were one baronet, six magistrates, one general, three colonels, one major, thirty noncommissioned officers and men, four clergymen, seven Tichborne tenants, and sixteen household servants.

The defense brought only seventeen witnesses, but they, along with the testimony of the claimant himself, carried the day. Sir John had exposed so many contradictions, distortions, absences of fact, and outright blunders that it became clear to the jury that the hundred witnesses who had testified for the claimant were mistaken. The claimant was immediately taken into custody, charged with three counts of perjury, and remanded for criminal trial.

Between the two trials the truth about the claimant came slowly to light. His real name was Arthur Orton. He was the son of a butcher residing at 69 High Street, Wapping, and had practiced the butcher's trade himself in South America and Australia. After reading one of Lady Tichborne's advertisements in an Australian paper, as a lark he had begun calling himself Roger Tichborne and assuming what he imagined were noble airs (mostly copied from music hall sketches).

At the criminal trial Orton was sentenced to fourteen years in prison and served nearly eleven—a couple of years were knocked off for good behavior. When he got out, he wrote up and sold an account of his "true story." But he didn't get much for it—by then the public had lost interest.

But Orton had done an impressive, almost credible job of fooling people. If he had not gone to court to try to claim the estate, he might well have remained a Tichborne for the rest of his life. The family might even have supported him in some meager manner if he had agreed to stay away. Consider that more than a hundred people who had known the real Roger swore in open court that Orton was indeed Roger returned. He took advantage of his time in England to visit all the places Roger had been, to practice Roger's handwriting, to learn to walk and talk like a gentleman, and even to study French, though he could never manage to get the hang of it.

The time, expense, and notoriety of the trial, as well as Orton's near success at pulling it off, clearly showed the British authorities that a better means of identifying people needed to be found.

An infallible means of identifying those who, for whatever reason, were in the hands of the police was the goal, and the

The New York police persuade a man to have his picture taken, 1906.

invention of photography brought it one step closer. As early as 1854, daguerreotypes of criminals were being made in Switzerland. A delicate process that uses wet plates and exposure times of up to two minutes, the procedure must have been trying on both the authorities and the suspects. In the 1860s the Paris commissioner of police, Lêon Renault, set up the first photographic studio intended specifically for police use. By the 1880s, after the development of more efficient dry-plate processes, photography became an essential police tool. Pictures of criminals were taken upon either apprehension or conviction, depending on the local laws. Photographs of habitual criminals were gathered together in large "rogues' galleries" that were regularly studied by the police. A detective's ability to memorize the faces of habitual criminals in his district was his most useful tool in crime detection and prevention.

Unfortunately, photographs did not prove as reliable as had been assumed. First offenders who happened to resemble habitual criminals were given harsher sentences than they deserved. And the reverse also occurred: in 1888 a convict in Manchester Prison serving a light sentence as a first offender murdered a warder; in the subsequent investigation it was found that he was a known criminal whose appearance had changed sufficiently to fool the camera.

What was needed was a foolproof means of telling one person from another—a measure of a man that would differentiate him from all other men. Ideally this would be something that could be easily measured, that would not change over time, and that could be codified and retrieved when needed. It wasn't until the late 1870s that such a way was devised.

4 : The Size of the Head, the Shape of the Ears

THE FIRST MAJOR STEP toward a scientifically rigorous means of criminal identification was taken by a civilian employee of the Paris police by the name of Alphonse Bertillon. In 1878, Bertillon was a records clerk in the Paris prefecture of police. His job made him acutely aware of the difficulty of identifying repeat offenders, and his family background suggested a possible solution.

Bertillon was born in Paris in 1853. Although this was an era in which scientists were making rapid forward strides, it was still possible for a generalist to contribute usefully to several fields of knowledge. Bertillon's father, Louis-Adolphe, a descendant of a family of vinegar makers from Dijon, was a medical doctor as well as an anthropologist and statistician. The anthropologist Paul Broca, whose researches into the human brain first located the speech center, was a regular visitor at the Bertillon house and a sometime colleague of Dr. Bertillon.

From his childhood association with Broca and his father, the young police clerk had an intimate knowledge of the human skull. He first saw a way to apply anthropological studies to the work of the police when he heard of a theory of Lambert Adolphe Jacques Quetelet (1796–1874).

Quetelet was one of those polymathic geniuses who seem so common to the eighteenth and nineteenth centuries. A noted astronomer and mathematician, he was a founder of modern statistics, a method he developed in order to reduce errors in the measurement of stars. He was also an important innovator in the fields of meteorology, demography, sociology, and criminology, and was among the first to apply statistical methods to the study of human behavior. His most influential book was his 1835 *Treatise on Man* (or, as the French called it, *Sur l'homme et le développement de ses facultés, ou Essai de physique sociale).*

Quetelet believed that the differences among individuals could be quantified, and that they fell along a normal distribution curve. His studies were among the earliest to show a relationship between crime factors such as age, gender, poverty, education, and alcohol consumption. He carried his theories of measurement of differences from the social realm to the physiological, observing that no two people have exactly the same body measurements.

Bertillon saw that if Quetelet was correct, it would be necessary to compare the measurements of two people only at a limited number of points in order to tell them apart. And if one kept a record of these measurements, one could identify a person with certainty if one ever had the occasion to measure him again. This seemed to Bertillon to have a direct application to the problem of identifying repeat offenders.

Bertillon worked out a system he called anthropometrics, but which his French compatriots came to call Bertillonage. Its three basic tenets were as follows:

1. That after about the age of twenty, the underlying bone structure of the human body is fixed.

2. That by using the right measurements one could differentiate every member of the human race.

3. That these measurements might be made with sufficient precision without undue difficulty.

In 1879, Bertillon took his idea to Prefect of Police Louis Andrieux, but the prefect's response was less than enthusiastic: he threatened Bertillon with immediate dismissal if he continued to waste time on such a project. Fortunately for the future of anthropometrics, in 1882 Andrieux was replaced by Jean Camescasse, who gave Bertillon permission to conduct a three-month trial of his system. Two months and one week later, on February 24, 1883, Bertillon successfully identified his first recidivist. In 1883 he identified 43 recidivists, and in 1884 the number rose to 241. These successes were so impressive that a new service of judicial identity was formed, with Bertillon placed at its head.

The system that Bertillon developed had three components:

1. Measurements of the body: the height of the subject standing; his reach (the distance from fingertip to fingertip when the arms are outstretched); and the height of the subject when seated.

2. Measurements of the head: its maximum length from front to back; and the length and diameter of the right ear from upper left to lower right.

3. Measurement of the limbs: the length of the left foot; the lengths of the left-middle and little fingers; length of the left forearm from the elbow to the tip of the middle finger. The left side was chosen for limb measurements because it was believed to be less subject to changes from the stresses of physical work.

Bertillon constructed special stools and calipers in order to standardize the measuring process. Nonetheless the person taking the measurements had to be trained to do it correctly or the results would be anything but uniform. Even in the best of situations the measurements would vary from one taker to another. Bertillon was forced to publish a table of tolerances to allow for such errors. Another difficulty was the large number of criminals under twenty—their anthropometric measurements were still subject to the changes of maturation.

These anthropometric records would be useless without a
specific record for comparison. The French criminalist Edmond
Locard described the system that was devised thus:

> The mass of a service's files is divided into three groups,
> according to whether the length of the head is large,
> medium, or small. Each of the groups is divided into three
> subgroups according to the breadth of the head. Then each
> subgroup in its turn is subdivided into three classes by the
> length of the middle finger. And each class of middle fingers
> into three categories by the length of the ring finger.

In 1896, Ida Tarbell, a well-known and accomplished American journalist, wrote a long article on Bertillonage for *McClure's
Magazine*. When she interviewed Bertillon at his office in Paris,
Miss Tarbell found him to be "a tall man of slightly haughty
bearing. He had a grave face, of long regular lines; a dark, almost
melancholy eye, with the slight contraction of the lids peculiar to
serious students, and a nervous trick of knitting his brow." His
contribution to society, and particularly to forensic science, was
unparalleled, she wrote. Bertillon was

> the man who has so mastered the peculiarities of the human
> anatomy and so classified and organized his observations,
> that the prisoner who passes through his hands is subjected
> to measurements and descriptions that leave him forever
> "spotted." He may efface his tattooing, compress his chest,
> dye his hair, extract his teeth, scar his body, dissimulate his
> height. It is useless. The record against him is unfailing. He
> cannot pass the Bertillon archives without recognition; and,
> if he is at large, the relentless record may be made to follow
> him into every corner of the globe where there is a printing
> press, and every man who reads may become a detective
> furnished with information which will establish his identity.
> He is never again safe.

A Bertillon form used by the New Orleans Police Department, 1907.

After a short while Bertillon found it expedient to add a fourth element to his system: a pair of photographs, one in full and one in profile, taken with a special camera. Attached to the photography was a written description, a *portrait parlé*, that included items such as birthmarks, scars, tattoos, and noticeable deformities.

For a contemporary account of how the system worked, I again turn to Ida Tarbell, who had the privilege of watching as Bertillon's anthropometrically trained officers measured a thief:

> "Call the prisoner," said my guide, M. David of the service, and immediately the guard brought in a short, rather

stout man, clad only in undershirt and trousers. His feet
were bare. His face was not at all disagreeable, and his eyes
were bright and dark. He seemed to be perfectly indifferent
to what awaited him, and gave his name and country
without hesitation.

"He has been arrested for stealing rabbits at Robinson,"
said my guide. "Our business is to find out if he has ever
been up before. We'll make the observations together, and
you may record them on this card," handing me a piece of
card-board with many peculiar divisions and sub-divisions
marked on it.

"*Observations Anthropometriques*" was the
introductory heading, and "height" the first division.
The prisoner was directed to place himself against a high
measuring board, bearing at the side a scale. A flat board
was placed across the top of his head, and the height it
marked noted.

"Five hundred and fifty-eight," said my guide. As I made
my entry, a clerk in a high desk at the side repeated the
number and wrote it in the book before him. "Of course,"
said M. David, "it is understood that it is one metre, 55.8
centimetres" (five feet, 1.34 inches).

Without changing his position, the arms of the prisoner
were stretched at full length, and the third measure taken—
one metre, fifty-nine centimetres (five feet, 2.6 inches). The
second measure, the curvature of the spine, is rarely taken.
The fourth, height of the trunk, followed—eighty-three
centimetres (two feet, 8.68 inches).

The next step was a little more complicated. The subject
was ordered to sit down, and a jointed compass, furnished
with a semi-circular scale divided into millimeters, was
applied to his head, one foot being braced against the root

of the nose, and the other moved over back of the skull, in search of the point of greatest depth. . . . After three trials the greatest depth was found, and 19.2 centimeters (7.56 inches) read out. In the same way the width was taken, 17.3 (6.8 inches), and then followed measurements of the ear.

"These measurements of the head," said my guide, "are of extreme importance, because so sure. A tricky subject may expand his chest or shrink his stature, but he cannot add to or subtract from the length and breadth of the skull. And now for the foot."

The prisoner was told to step upon a stool, and throw back the right leg in such a way that the entire weight should come upon the left foot. The measuring of the foot was followed by that of the left middle and little fingers and of the left forearm. "All good measures," observed my conductor; "for the rule rests against the bones, and no dissimulation is possible on the part of the subject, and the chance for error on the part of the operator is little. And now for the eyes."

The man was placed in a strong, full light, and told to regard the operator in the face. The latter then raised the left eyelid slightly, and seemed to be making mental notes of what he saw.

"But the eye changes," I objected. "That man's eye ought to be darker now, under the excitement of this examination."

"False notion, that of the eye changing so much," said my guide. "It is the ground of the iris which is affected chiefly by the light; and we do not base our classification on that. Here are the notes."

"Class 3–4: aureole, radiant, of medium chestnut; periphery, of medium greenish yellow; two circles equal."

"But where do you get all that information?" I queried.
The gist of the answer I received was as follows:

The color of the eye is the result of the fusion of two
elements, the shade of the ground of the iris and that of
the aureole which surrounds the pupil. The usual method
of classifying eyes in the past has been to regard them at a
distance of three or four feet, and to mark the result of the
fusion of the two elements. Eyes thus studied are classified
as blue, brown green, and gray, or as dark, medium, light.
But there is little precision in this method. M. Bertillon
resolved to study the eye close ahand, and to analyze each
of the elements. He found that the ground of the iris is
rarely decided in shade, varying from a sky blue to a slate
blue, and changing according to the intensity of the light
and is, therefore, of little service in an exact description.
The pigment of the aureole around the pupil is, however,
more pronounced in color, and less variable in the light, and
therefore better capable of serving as a basis of classification.
By means of it the eye can be separated into seven
sufficiently distinct classes.

(1). Pale, or without pigment; that is, an eye in which the
aureole is absent or very insignificant, and in which the iris
is marked by whitish striae.

(2). Yellow aureole.

(3). Orange aureole.

(4). Chestnut aureole.

(5). Maroon aureole in a circle or disk around the pupil.

(6). Maroon aureole covering the iris irregularly.

(7). Maroon aureole covering the entire iris.

Each of these divisions may be further divided into light,
medium, dark, according to the shade. The sub-divisions
approach closely sometimes; thus an eye may appear to one
person as a dark orange, which to another will seem light
chestnut.

When there is a doubt, the two classes are marked: thus, in the case of our rabbit man, the class was three or four. After the class is decided, the arrangement of the aureole is noted. Is it a solid, definite circle? Does it send off short rays? Do the rays touch the periphery? Do they cover the iris? Is it mottled by a different shade? All the mosaics, the festoons, the lace-like drapings of the aureole are noted. In the same way the color and the arrangement of the periphery of the iris are described. If there are striking peculiarities, they are added to the list.

"There are still two classes of points to be taken," said M. David, "the descriptions and the special marks and scars; but we have now all that is essential. You may go," to the prisoner.

The *portrait parlé* was also meant to stand alone as a verbal identification system. From his written description, a policeman trained in the *portrait parlé* could recognize a person in a crowd. But with hundreds of possible subdivisions and fine details, it took time and intelligence to master.

As originally devised by Bertillon, there were four major sections to the *portrait*:

1. A determination of the color of the left eye, hair, beard, and skin.

2. A morphological description of the various parts of the head, with emphasis on the right ear.

3. General considerations—body shape, carriage, voice, language or accent, clothing, apparent social standing, etc.

4. Indelible markings: scars, tattoos, birthmarks, and the like.

The recorder was to note peculiarities such as freckles and pockmarks. Hair could be light blond, blond, dark blond, brown, black, red, white, mixed grey, or grey. Baldness was also noted and described as either frontal, occipital, top-of-head, or full. Eyes were either blue, grey, maroon, yellow, light brown, brown, or dark brown. There were also various peculiarities of the eye—extremely bloodshot, eyes of two different colors, and *arcus senilis*, a white ring around the edge of the cornea.

And while the shape of the nose was of prime importance, it was by the ear that one might truly know the man. Barring accident, its shape is changeless from birth to death, making it the surest means of confirming identification by photograph. In its position and its angle on the head, the ear varies from person to person. It may take one of four general shapes: round, oval, rectangular, or triangular. Unlike Gaul, it is divided into six parts, the *helix*, the *antihelix*, the *tragus*, the *antitragus*, the *lobule*, and the *concha*. The helix is further subdivided: it may be either folded or flat and can form a variety of angles. The other parts also have distinct peculiarities that when added together account for tens of thousands of possible variations—a large enough number for preliminary elimination, but not large enough for positive identification.

Back to Miss Tarbell for a contemporary description of how the individual cards were classified and subsequently retrieved:

> "To classify, we select the measures which are the surest
> [said her guide]; that is, those which do not vary with
> age; which the individual cannot change; which are the
> most valuable from one person to another, and which the
> operators make the fewest errors in taking. Experience has
> taught us that these are: (1) the length of the head; (2) the
> breadth of the head; (3) the middle finger; (4) the foot; (5)
> the fore-arm." . . .

From the full card cabinet with ninety thousand cards, each representing one man or woman in the criminal catalog, the job was to find which card matched the man now being measured.

With each division—the width of the head, the length of the middle finger to the first joint, the length of the foot, the length of the forearm from elbow to fingertip—the number of possibilities was reduced until it was down to a mere four hundred. Then, with the height and the length of the little finger added, the pack was winnowed down to about sixty. Next came the color of the eye, and the sixty was reduced to a dozen.

And last—distinguishing marks. A mole on his left arm? The prisoner rolled up his sleeve and there it was. Another mole on the wrist? There it was. They had identified their man.

At the time, the recording of distinctive marks—scars, moles, tattoos, and the like—was one of the strong points of Bertillon's *portrait parlé*. But advances in surgery have made such things untrustworthy over time.

The *portrait parlé* was modified over the years into a simpler and more-or-less uniform identification form that was used by most police forces until the mid-twentieth century. One of Bertillon's students, Harry Ashton-Wolfe, describes the advantages of the *portrait parlé* with enthusiasm:

> One of the reasons why the layman cannot memorize a face in words is that he has no vocabulary to suit the need, and it is this vocabulary which was one of the first things the experts created. When we can describe a man as medium-tall, muscular, and corpulent; brachycephalic (round-headed) with low forehead; short, straight black hair coming to a point between bushy eyebrows; clean-shaven, with large ears set at right angles to the skull; eyes black, deep-set, small, and mobile; nose flat, with wide nostrils; fleshy lips; angry red bullet scar on left cheek; bulldog chin; abnormal canine teeth—we already have a mental picture of a simian, criminal type which would fix his unpleasant personality on our memory.

Bertillon advanced the science of photographic identification by a system he called metric photography, in which the camera, the lens, the distance of the camera from the subject, and the chair the subject sat in were standardized. He suggested a standardized reduction ratio of 1:7 for all photographs, so that the relative sizes of the subjects' heads were immediately apparent. He also pointed out something that should not have needed mentioning—that the photographic negative should not be retouched to remove scars or blemishes from the subject's face.

Bertillon worked out a method of improving the use of the camera in photographing crime scenes. He used standardized lenses, he took his pictures from standardized heights above the floor, and he printed on special paper imprinted with either an indoor or an outdoor grid, the outdoor grid allowing for distances to infinity.

Bertillon also made lesser-known but nonetheless important advances—the "galvano-plastic" method of preserving footprints found at the scene of the crime, for instance. Perhaps his only notable failure was his attempt to develop a system of handwriting analysis. Unfortunately he allowed himself to believe that he had succeeded. His aim was to identify forgeries or disguised handwriting and determine their author. Bertillon had studied the problem and developed some theories. His espousal of his incorrect ideas thrust him into the middle of an affair that became one of France's greatest political and social crises.

On October 15, 1894, Captain Albert Dreyfus of the French army was arrested on a charge of treason. He was alleged to have written a letter (known throughout France as the *bordereau* [memorandum]) in which he offered to pass on secret information to the Germans. In fact Dreyfus was completely innocent. But he was a Jew at a time when the French army's general staff was rabidly anti-Semitic.

The prosecutors asked Bertillon, the renowned head of the Paris prefecture's Service of Judiciary Identity, to evaluate the *bordereau.* After a careful examination, Bertillon declared it a "self-forgery," that is, that Dreyfus had attempted to imitate his own handwriting. He had done this presumably in order to argue that the document was a forgery in the event he were caught. Why he did not simply imitate someone else's handwriting Bertillon did not say. Somewhere in the course of his investigation, Bertillon developed an *idée fixe* of Dreyfus's guilt that warped his neutrality and thus his judgment.

Bertillon prepared a document showing where Dreyfus had employed a number of "deviations, shifts, or displacements" and had traced some of the words seven or eight times before writing them. According to Bertillon's investigations, Dreyfus had even borrowed the shapes of certain letters from other members of his family. Jean Casimir-Périer, the president of the republic, given a demonstration of Bertillon's conclusions, described Bertillon to a friend as "not merely bizarre, but completely insane, given to an extraordinary and cabalistic madness."

Bertillon told the army that "The proof is there and is irrefutable. From the very first day, you knew my opinion. It is now absolute, complete, and admitting of no reservation." The army wanted to believe. Dreyfus was convicted and sent to Devil's Island. After two more trials and the fall of two governments, the identity of the real traitor, Commandant Count Ferdinand Walsin-Esterhazy, was revealed. Dreyfus was exonerated and reinstated in the army in time to fight in World War I. The affair blighted the reputation of Bertillon and cast a shadow over his truly notable achievements.

While Bertillon identified criminals by measurements of their heads, the Italian criminologist Cesare Lombroso (1836–1909), director of an insane asylum at Pesaro, Italy, took this system one step further. Misunderstanding Darwin's theory of evolution, Lombroso decided that criminals differed from their more law-abiding brethren because of traits that were either atavistic or degenerative. Where the atavistic traits were throwbacks to an earlier stage of evolution, the degenerate ones had progressed in the wrong direction along the evolutionary line. Both types exhibited physical anomalies that made it possible to recognize them even before they committed the criminal acts they were doomed by their natures to perform. The atavists could be recognized by their subhuman characteristics—large jaw, a bulging brow, high cheekbones, and extra-long arms. According to Lombroso, these were "reminiscent of apes and lower primates, which occur in the more simian fossil men, and are to some extent retained in modern savages."

The degenerates could be known by their congenital weaknesses and by their slack-jawed, unintelligent appearance. Tattooing, unless extremely minor, was another symptom of degeneracy. Lombroso used Bertillon's anthropometric techniques to collect measurements of his "criminal types," and believed he had found a meaningful correlation.

Many followers of Bertillon returned the compliment by firmly embracing Lombroso's theories. As Harry Ashton-Wolfe explained:

> Criminal faces, as the police and the laboratory experts
> understand them, are divided into three sections: those
> of degenerates or throw-backs, which vary little during
> their lifetime; those stamped with the evil characteristics
> which a career of crime inevitably evolves through constant
> association with others of the species and a frequent sojourn
> in penal establishments; and, finally, the faces which reveal
> Darwinian deformation and asymmetry of the features
> which, in many instances, may be merely some single
> strikingly abnormal development due to hereditary criminal
> tendencies.

For a while in the late 1800s Lombroso reigned supreme. Quoted everywhere, his doctrines were embraced by the criminal courts of Europe. "*L'uomo delinquente*" (the born criminal), believed to have more savage inclinations than the most savage ape, had better be put safely away for his own good and the good of humanity. Many who were accused of comparatively minor crimes received substantially longer sentences than their brethren because they had thick brows or long arms.

Bertillon himself wasn't so sure, telling Ida Tarbell:

> "No; I do not feel convinced that it is the lack of symmetry in the visage, or the size of the orbit, or the shape of the jaw, which make a man an evil-doer. A certain characteristic may incapacitate him for fulfilling his duties, thus thrusting him down in the struggle for life, and he becomes a criminal because he is down. Lombroso, for example, might say that, since there is a spot on the eye of the majority of criminals, therefore the spot on the eye indicates a tendency to crime; not at all. The spot is a sign of defective vision, and the man who does not see well is a poorer workman than he who has a strong, keen eyesight. He falls behind in his trade, loses heart, takes to bad ways, and turns up in the criminal ranks. It was not the spot on his eye which made him a criminal; it only prevented his having an equal chance with his comrades. The same thing is true of other so-called criminal signs. One needs to exercise great discretion in making anthropological deductions. Nevertheless, there is no doubt but that our archives have much to tell on all questions of criminal anthropology."

But Bertillon, who seemed to have an ear fixation, averred that an infallible sign of either atavism or degeneration was a "striking asymmetry and malformation of the ears." Lombroso's later contention that genius was another form of degeneracy brought him into disrepute, especially among those who believed themselves to be geniuses.

The form of *portrait parlé* used in the United States before World
War II, as recorded in Söderman and O'Connell's *Modern Criminal Investigation* (1935), looked like this:

NAME ...
SEX ...
COLOR ...
NATIONALITY ...
OCCUPATION ...
AGE ...
HEIGHT ...
WEIGHT ...
BUILD—Large; stout or very stout, medium; slim; stooped
    or square-shouldered; stocky.
COMPLEXION—Florid; sallow; pale; fair; dark.
HAIR—Color; thick or thin; bald or partly bald; curly;
    kinky; wavy; how cut or parted; style of hairdress.
EYES—Color of the iris; eyes bulgy or small; any
    peculiarities.
EYEBROWS—Slanting, up or down; bushy or meeting;
    arched, wavy, horizontal; as to texture, strong; thin;
    short or long-haired; penciled.
NOSE—Small or large; pug, hooked, straight, flat.
WHISKERS—Color; Vandyke; straight; rounded; chin
    whiskers; goatee; side whiskers.
MUSTACHE—Color; short; stubby; long; pointed ends;
    turned-up ends; Kaiser style.
CHIN—Small, large; square; dimpled; double; flat; arched.
FACE—Long; round; square; peg-top; fat; thin.
NECK—Long; short; thick; thin; folds in back of neck;
    puffed neck; prominent Adam's apple.
LIPS—Thick; thin; puffy; drooping lower; upturned upper.

MOUTH—Large; small; drooping or upturned at corners; open; crooked; distorted during speech or laughter; contorted.

HEAD—Posture of—bent forward; turned sideways; to left or right; inclined backwards or to left or right.

EARS—Small; large; close to or projecting out from head; pierced.

FOREHEAD—High; low; sloping; bulging; straight; receding.

DISTINCTIVE MARKS—Scars; moles; missing fingers or teeth; gold teeth; tattoo marks; lameness; bow legs; pigeon toes; knock-knees; cauliflower ears; pockmarked; flat feet; nicotine fingers; freckles; birthmarks.

PECULIARITIES—Twitching of features; rapid or slow gait; long or short steps; wearing of eyeglasses; carrying a cane; stuttering; gruff or effeminate voice.

CLOTHES—Hat and shoes—color and style; suit—color, cut, maker's name; shirt and collar—style and color; tie—style and color; dressed neatly or carelessly.

JEWELRY—Kind of; where worn.

WHERE LIKELY TO BE FOUND—Residence; former residences; places frequented or hangouts; where employed; residences of relatives, etc.

PERSONAL ASSOCIATES—Friends who would be most likely to know of the movements or whereabouts of the person wanted, or with whom he would be most likely to communicate.

HABITS—Heavy drinker or smoker; drug addiction; gambler; frequenter of pool parlors; dance halls; cabarets; baseball games; resorts, etc.

HOW HE LEFT THE SCENE OF THE CRIME—Running; walking; by vehicle; direction taken.

5 : Dabs

The examination of finger-prints is no
easy matter. It is therefore above all
necessary that good and true impressions
should be taken which can be kept
and compared with others.
—Hans Gross

AN ANCIENT BABLYONIAN tablet in the British Museum holds the cuneiform record of an officer's testimony. In it he tells of being sent by his superior to arrest the defendant, confiscate his property, and take his fingerprints. Because many tablets from this period detail business transactions that were concluded with a fingerprint impressed into wet clay, it is reasonable to assume that 4,500 years ago the Babylonians understood the uniqueness of individual fingerprints.

Some fingerprint fanatics even believe that the seventeenth verse of the Christian apostle Paul's second letter to the Thessalonians ("The salutation of Paul with mine own hand, which is the token in every epistle: so I write") is evidence that Paul signed the letter with his fingerprint.

In the third century B.C. a Chinese official pressed his thumb into one side of a clay seal and wrote his name on the other, leading us to the inescapable conclusion that the print's purpose was one of identification. Thus two thousand years ago someone knew of the immutability of fingerprints. They have shown up in the designs of pottery far older than this. But there is no way to know if these impressions were merely accidents, a design feature, or the potter's method of signing his work.

The bureaucrats of China's Tang dynasty (618–907 A.D.) seem to have been aware of the individuality of fingerprints. Kai Kung-yen, an author of the period, writes of the wooden tablets used for contracts centuries earlier, before the invention of rice paper. As he explains, at one time identical notches were cut in the sides of two tablets, each with the contracts written on them. One tablet was then given to each party. "The significance of these notches," Kai Kung-yen explained, "is the same as that of the finger prints (*hua chi*) of the present time."

Bernard Laufer's *History of the Finger-Print System* contains an account of the Arab merchant Soleiman. In 851 A.D. he wrote:

> The Chinese respect justice in their transactions and in judicial proceedings. When anybody lends a sum of money to another, he writes a bill to this effect. The debtor, on his part, drafts a bill and marks it with two of his fingers united, the middle finger and the index. The two bills are joined together and folded, some characters being written on the spot separating them; then, they are unfolded and the lender receives the bill by which the borrower acknowledges his debt.

In the thirteenth-century Chinese novel *Shui hu chuan* a husband smears his hand with red ink and presses it onto a divorce document to make it legal. The eighth-century Japanese had a similar practice—there are documents in Japanese archives that bear the handprints of many a long-dead Mikado. The word *tegata* (hand stamp) survives as a reminder.

A Japanese official presses his fingerprints onto a document, circa 1300 A.D.

A thousand years ago in Tibet, impressions of hands and feet had a religious meaning. As Bernard Laufer explains:

> These notions were apparently derived with Buddhism from India. In the Himalayan region of southern Tibet the pious believers are still shown foot imprints left by the famous mystic, ascetic, and poet, Milaraspa (1038–1122). . . . By "traces of snowshoes" is still designated a boulder on which

he performed a dance and left the traces of his feet and staff, and the fairies attending on the solitary recluse marked the rocks with their footprints.

In the fourteenth century, Tamerlane (or Timur, as he was known in his own Mongol language) conquered much of what is now Turkey and Iran and founded the brief-lived Timurid Empire. To control his conquered lands, "The officers of the conqueror's army were appointed to the charge of the different provinces and cities which had been subdued, and on their commissions, instead of a seal, an impression of a red hand was stamped; a Tartar usage, that marked the manner in which the territories had been taken, as well as that in which it was intended they should be governed."

The Roman barrister and teacher Quintilian, or, as the Romans called him, Marcus Fabius Quintilianus, was the first person we know of to use fingerprints or, more correctly, handprints, as evidence at trial. In about 72 A.D., Quintilian defended a blind teenage boy accused of murdering his father. The evidence? Bloody handprints leading from the father's bedroom and back along the corridor to the boy's room. The boy—we do not know his name—had saved his father's life two years earlier when their house caught fire. In that earlier incident, he had rushed back into the house to save his mother but had been too late. The attempt blinded him.

Did he blame his father for his mother's death? Had the blame festered and grown as the months passed? Had he, in a moment of unbearable grief, slipped into his father's room, slid a knife between his father's ribs, and neatly pierced his heart? Had he then staggered back to his own room, guiding himself down the corridor by putting his hands against the wall, thus leaving a damning trail of bloody handprints? The prosecution, brought by the father's second wife, the lad's stepmother, thought so.

Unless, they suggested, he had merely murdered his father for his inheritance.

"A nice theory," Quintilian said, "but it postulates improbable behavior on the part of the boy and *impossible* behavior on the part of the blood."

He described how the boy would have had to move swiftly down the hall carrying a knife. The blind boy would do this without bumping into anything, without knowing if a lamp burned or if someone, perhaps a slave, waited somewhere along his path, watching him as he passed. He would then have had to enter his father's bedroom without being able to see if his father or stepmother were awake; he would have approached the bed without stumbling and then thrust a knife faultlessly between his father's third and fourth ribs and into his heart. Then the lad would have soaked his hands in his father's blood before stumbling back to his own room, where he was found sleeping, innocent of bloodstains, the next morning.

And the handprints? Didn't a clear line of bloody impressions along the corridor leading from the death chamber to the lad's bedroom prove the lad's guilt?

"Nonsense," said Quintilian. Just the opposite. They were proof the boy was innocent.

"Blood," he told the jury, "does not behave that way. Surely you veterans of the legions of Rome should understand the ways of shed blood. All the bloody handprints were even and clear. If the boy had indeed staggered down the hall, resting his blood-soaked hand against the wall as he went, the prints would have become fainter and fainter as he went. To achieve a row of clear prints, someone had to keep his or her hand wet with blood by dipping it again and again in the corpse's wound. Someone who wanted to blame the boy. Someone who had eyes to see where to insert the knife blade. Someone who would inherit a fortune if, and only if, the boy were convicted of patricide."

The boy was found innocent. According to the record, the stepmother subsequently confessed.

The world waited almost two millennia before fingerprints, or, as the British police called them, "dabs," would figure again in a murder trial. Two facts had to be established before fingerprints could be used for positive identification: that they are unique—that no two people, even identical twins, have the same fingerprints; and that a person carries the fingerprints he is born with unchanged throughout his life. To make fingerprints truly useful, it was also necessary to devise a way to locate a specific fingerprint out of the myriad in the file.

In 1684, Nehemiah Grew, an English doctor and a fellow of the Royal Society, published a lecture in which he commented on the ridge patterns on fingertips. If it occurred to him that these could be used for identification purposes, he did not mention it.

Two years later, Marcello Malpighi (1628–1694), a professor of anatomy at the University of Bologna who gave his name to the Malpighi layer, one of the subdivisions of human skin, published *De Externo Tactus Organo* (*Concerning the External Organs of Feeling*). In it he described the ridged pattern of the skin of the fingers and palm. Again, he did not suggest that anything useful could be done with the information.

In 1818 the artist and engraver Thomas Bewick (1753–1828) published an edition of *Aesop's Fables* which he illustrated with his own woodcuts. On the title page he placed a woodcut of one of his own fingerprints over the legend, "Thomas Bewick—his mark." We can assume that Bewick was at least artistically aware of the uniqueness of fingerprints.

Johannes Evangelist Purkinje (1787–1869), a professor of anatomy and physiology at Breslau University, was the first person to attempt to examine fingerprints and classify them by type. His name is associated with many medical discoveries—Purkinje's cells, fibers, networks, vesicles, and more. In a thesis written in

1823 for his doctor of medicine degree at the University of Breslau, he divided fingerprints into nine groupings. As he said in the thesis: "After numerous observations, I have thus far met with nine principal varieties of curvature according to which tactile furrows, or furrows susceptible to touch, are disposed upon the inner surface of the last phalanx of the fingers."

In 1858, William James Herschel, then employed by the East India Company in Bengal, India, was in Jungipur on the upper reaches of the Hooghly River. He was in the process of drawing up a contract for road-building materials with a supplier named Rajyadhar Konai, a man who was known to take such agreements lightly. In order to impress upon him the importance of the contract, Herschel had Konai place his handprint on the reverse of the document. Two years later Herschel, then a magistrate at Nuddea, near Calcutta, was charged with seeing that the many natives who were destitute in the aftermath of the Great Mutiny received their government pensions. As most of the natives could neither read nor write and thus could not sign anything, the possibilities for fraud were endless. The criminal element was quick to exploit these possibilities.

Although he had taken Konai's handprint as a talisman of the seriousness of keeping one's word, Herschel realized that fingerprints could also serve as a positive means of identification. Once he began placing the pensioners' thumbprints on their receipts, the number of fraudulent claims fell dramatically.

For the next two decades, Herschel continued his study of fingerprints. He became convinced that they did not change with age, and that no two were alike. In 1877, in a letter to the inspector general of the prison system of Bengal, he outlined his researches: "I have taken thousands [of fingerprints] now in the course of the last twenty years, and I am prepared to answer for the identity of every person whose sign manual I can now produce if I am confronted with him." But the inspector general refused Herschel permission to try out his system, even on a small scale in a local prison. Disillusioned and in poor health, Herschel returned to England in 1879.

While Herschel experimented in India, Dr. Henry Faulds (1843–1930), a Scot who served as a resident physician at the Tsukiji Hospital in Tokyo, became fascinated with fingerprints. His interest began when he noted the impressions of potters' fingers on specimens of prehistoric Japanese pottery. He began a systematic study of fingerprints, determining that each of the prints he collected was unique and that an individual's prints did not change over the course of a lifetime. He found that the best medium for transferring the prints was a thin film of printer's ink on which the finger was rolled before being rolled again onto a card—a method still in general use today. In 1879, Faulds was able to use his developing science to aid the local police, thus becoming the first person to solve a crime with fingerprint evidence.

While inquiring into a burglary near Faulds's home, police investigators noticed some grimy fingerprints on a wall of the house. Knowing of Faulds's fascination with fingerprints, they invited him to examine the marks. At about the same time they also arrested a suspect. Faulds took the suspect's fingerprints and, after comparing them to the marks on the wall, declared him to be innocent. When a second suspect was apprehended a few days later, Faulds found that his prints matched the ones on the wall. A short while later Faulds helped the police again by lifting the fingerprints from a mug and comparing them with those of the suspect. They matched, and another burglar was caught.

A year later, on October 28, 1880, a letter from Faulds appeared in the British journal *Nature*. In it he outlined his discoveries and proposed the establishment of a scientific method of fingerprint identification. He also discussed his own experience in using fingerprints to solve crimes, saying in part, "When bloody fingermarks or impressions on clay, glass, etc., exist, they may lead to the scientific identification of criminals."

A reply from Herschel, in which he related his own experience in India, appeared in the next issue of *Nature*. Thus began a feud between the two men over who could claim priority in the use of fingerprints for identification.

In 1882 in the New Mexico Territory of the United States, a government geologist named Gilbert Thompson began using his own thumbprint on pay orders and requisitions in an effort to prevent forgeries. Where he came up with the idea is not known. A year later Mark Twain published *Life on the Mississippi*, a memoir in which he recalls the solution of a crime by the use of fingerprints. "When I was a youth," Twain wrote,

> I knew an old Frenchman who had been a prison keeper for thirty years, and he told me that there was one thing about a person which never changed, from the cradle to the grave— the lines in the ball of the thumb; and he said that these lines were never exactly alike in the thumbs of any two human beings. In these days, we photograph the new criminal, and hang his picture in the Rogues' Gallery for future reference; but that Frenchman, in his day, used to take a print of the ball of a new prisoner's thumb and put that away for future reference. He always said that pictures were no good—"The thumb's the only sure thing," said he; "you can't disguise that." And he used to prove his theory, too, on my friends and acquaintances; it always succeeded.

We don't know if Mark Twain considered the story fact or fiction, but his old Frenchman had his facts right. He knew as much about fingerprinting as any man of his day and more than most. Twain's story is illustrated with woodcuts of fingerprints, whether Twain's or the artist's we do not know.

Twain's fascination with fingerprints continued. Ten years later a major plot point in his book *Pudd'nhead Wilson* revolved around a courtroom identification based on fingerprint evidence.

In 1886, Faulds, then back in Britain, offered to set up a fingerprint bureau for Scotland Yard at his own expense. His offer was rejected. The authorities realized that neither Herschel nor Faulds had developed a classification method that allowed for later retrieval. Without such a method, the value of fingerprints for identification was extremely limited.

Enter the noted British scientist Sir Francis Galton (1822–1911). In the 1880s he became interested in dactyloscopy, as the study of fingerprints had become known. Galton, a cousin of Charles Darwin, was a medical doctor as well as a geneticist and anthropologist. At the age of twenty-three he had led an expedition to the Sudan and southwest Africa; he later wrote *Meteorographica* (1863), a meteorological treatise in which he presented the modern method of weather mapping. He was sixty when his interest in anthropometry led him to the application of statistical methods in anthropology and heredity, an endeavor that resulted in the founding of the study of eugenics.

Sir Francis had for many years conducted statistical studies of human hereditary traits, an effort that gave him a working familiarity with anthropometric measurements. As a result of his study of inherited characteristics, he became proficient in Bertillon's system of body measurement. His work soon made him a recognized expert.

In 1888, Galton was asked to deliver a lecture on Bertillonage to the Royal Institution of Great Britain. In preparing his lecture, he focused on the problems of individual identification and became fascinated with the possibilities he saw in the use of fingerprints. As he later related in his book *Finger Prints*: "Wishing to treat the subject generally, and having a vague knowledge of the value sometimes assigned to finger marks, I made inquiries and was surprised to find how much had been done, and how much there remained to do before establishing their theoretical value and practical utility!"

Herschel kindly lent Galton his collection of fingerprint cards, and Galton spent the next three years establishing that finger-prints remained unchanged throughout a person's life and that it was theoretically possible to devise a method of classifying them. Using an extremely conservative estimate of the degree of vari-ability in a given fingerprint, Galton calculated the probability of two sets of prints matching to be 1 in 64 million. (The figure given today is much larger.) (Some popular articles on fingerprint-ing have claimed that identical twins have identical fingerprints. This is not so. In some cases the prints will correspond in primary and secondary classifications, which means that their cards will be filed in the same general area of the record section, but direct comparison of the two sets of prints will show that they are quite different.)

In *Finger Prints*, Galton declared that he had found finger-prints to be permanent and immutable:

> As there is no sign, except in one case, of change during
> any of these four intervals which together almost wholly
> cover the ordinary life of man (boyhood, early manhood,
> middle age, extreme old age), we are justified in inferring
> that between birth and death there is absolutely no change
> in, say, 699 out of 700 of the numerous characteristics of the
> markings of the fingers of the same person such as can be
> impressed by him wherever it is desirable to do so. Neither
> can there be any change after death up to the time when
> the skin perishes through decomposition: for example, the
> marks on the fingers of many Egyptian mummies and on the
> paws of monkeys still remain legible. Very good evidence
> and careful inquiry is thus seen to justify the popular idea

of the persistence of finger markings. There appear to be no bodily characteristics other than deep scars and tattoo marks comparable in their persistence to these markings; at the same time they are out of all proportion more numerous than any other measurable features. The dimensions of the limbs and body alter in the course of growth and decay; the color, quantity, and quality of the hair, the tint and quality of the skin, the number and set of the teeth, the expression of the features, the gestures, the handwriting, even the eye color, change after many years. There seems no persistence in the visible parts of the body except in these minute and hitherto disregarded ridges.

Galton developed a tentative approach to the problem of indexing fingerprints, an approach based upon whether the predominant pattern of each individual print was an arch, a loop, or a whorl. In 1893 the British Home Office established a committee headed by Charles Edward Troup to recommend a criminal identification system for use by Scotland Yard. The Troup Committee consulted with Galton and were impressed with the potential of the fingerprint identification method that he demonstrated in his laboratory. But Galton, the true scientist, explained to them that his system was too complex and unwieldy for use outside the laboratory. It would be some years, he said, before a reliable system could be perfected.

The Troup Committee therefore recommended that Bertillon's anthropometric system be adopted by Scotland Yard, but that it be supplemented by fingerprinting. This way, a library of fingerprints would be available when a comparative method was perfected. And this suggestion was approved.

In 1894, Galton, who was then seventy-two, passed the torch of dactyloscopy to Edward Henry (1850–1931). It was Henry who carried it to success.

The London-born son of a doctor, Edward Henry was a career civil servant who rose to the post of inspector general of police for the province of Bengal. While in India he had read Galton's *Finger Prints* and subsequently traveled to England especially to meet him. Galton, with his usual generosity, talked to Henry for hours and explained the problems of classification he had solved and those that remained. He loaded Henry down with as many fingerprint cards and pages of notes and examples as he could carry and sent him away.

Back in India, Henry worked on the problem for the next two years. He settled on five basic patterns for fingerprints: arches, tented arches, radial loops (slanting toward the thumb), ulnar loops (slanting away from the thumb), and whorls. He designated these by the letters A, T, R, U, and W. He analyzed them further by showing that a straight line that connected two specific locations on a print could be drawn and the number of ridges cut by that line then counted. These letters and numbers would produce a specific code for any fingerprint card that held the prints of all ten fingers. Any two investigators categorizing the same fingerprint card would come up with the same code.

The system was complex and required intensive effort to master, but it worked. Once it was in use, no felon could hope to hide his previous record or outstanding arrest warrants from the police.

In July 1897, after a one-year trial, the Indian government officially replaced Bertillonage with dactyloscopy. In 1901, Henry was called back to London to head Scotland Yard's Criminal Investigation Division (CID). As the assistant commissioner of police, he oversaw the discarding of anthropometry and the adoption of his own system of fingerprinting as the sole means of criminal identification. It quickly proved its usefulness, transforming the process of identifying criminals from a time-consuming task fraught with the possibility of error into a routine procedure that was as close to infallible as anything yet devised. If two fingerprint cards held matching prints, they had been made by the same person, and no question about it.

As Henry perfected his system, an Argentine detective was independently developing a fingerprint system of his own. Austrian-born Juan Vucetich (1858–1925) emigrated to Argentina at the age of twenty-six. Shortly after arriving, he joined the La Plata police department, where within five years he became head of the statistical bureau. In 1891 he set up an office of anthropometry for the identification bureau of the central police department in La Plata. As he worked with anthropometry, Vucetich became aware of its shortcomings and began to look for a better system. After reading an article on Galton's research in the May 1891 issue of *Revue Scientifique*, a French science magazine, Vucetich worked out his own system of four primary categories of fingerprints.

In 1892, while Vucetich was developing his system, a woman named Francesca Rojas, who lived in Moecochea, in the province of Buenos Aires, was found in her home in serious condition with stab wounds to her neck. Her two sons lay dead, both with their throats cut. Rojas accused a former lover who lived nearby of committing the crimes. The police, at the request of their superiors in La Plata, cut a bloodstained section of wood from the doorjamb and forwarded it to the identification bureau. There, a fingerprint was found in the bloodstain that proved to be that of Rojas herself. She confessed to the crime and went to prison.

Convinced of its value, for years Vucetich financed the development of fingerprint classification from his own meager salary. The authorities refused to consider switching from Bertillonage even when in one day Vucetich identified twenty-three criminals who had successfully fooled the Bertillon system. Finally, under a new police chief, in 1894 Argentina adopted the Vucetich system of classification. In 1896 Bertillonage was officially dropped.

In what may be a supreme irony, when Vucetich went to Paris
a few years later and met Bertillon, the Frenchman accused him
of using Bertillonage without permission or proper credit. Bertillon physically attacked him. What might Bertillon have done
had he known that Vucetich had abandoned anthropometry for
dactyloscopy?

In 1904, Vucetich explained his system in his book *Dacty-
loscopia Comparada*. In 1907 the French Académie des Sciences
judged it the best of all they examined, and by 1912 it was in use
by all the countries of South America.

Meanwhile Great Britain and most of the countries of Europe used the Galton-Henry system. France, Belgium, and Egypt
used an amalgam of the two systems, with France also retaining
anthropometry for the first half of the twentieth century. Even
Bertillon had by this time incorporated fingerprints into Bertillonage. Curiously, in 1902, during the first criminal prosecution
in France based on fingerprint evidence, it was Bertillon who
made the identification. A man named Joseph Reibel had been
strangled in a dentist's office on the Rue du Faubourg Sainte
Honoré in Paris, and a single bloody fingerprint had been left
on a glass shard at the scene. Bertillon photographed the print,
and, through intensive searching (at that time he had to methodically examine every card), he matched it to that of an ex-convict
named Scheffer. Several eyewitnesses were also able to identify
Scheffer from his photograph. When he was eventually apprehended in Marseille, he confessed.

In the United States the practice of fingerprinting and of main-
taining fingerprint files began in the major penitentiaries. Sing
Sing began using the Henry system in March 1903, and other
New York state prisons took it up shortly thereafter. The federal penitentiary in Leavenworth, Kansas, switched from anthropometry to fingerprinting toward the end of 1904. Officials there

were readily convinced because of an incident that had occurred there the year before. In 1903, as a new prisoner named Will West was being measured for his Bertillon record, Warden R. W. McClaughty asked why the man's record was being duplicated. West protested that he had never before been at Leavenworth or any other prison. The warden pulled the record of one William West, convict number 2626, from the files, and confirmed that it corresponded to the prisoner's Bertillon measurements. And the photograph looked just like him. When Will West persisted in his denials, the warden checked further and found that the William West of record 2626 was already a prisoner. The two men shared variants of the same name, looked the same down to the smallest detail, and shared identical Bertillon measurements. McClaughty had the prints of their left index fingers taken and compared, and was relieved to discover that in fact the two Wests were entirely different men.

The Bertillon measurements of William West, dated 9 September 1901, were: 19.7, 15.8, 12.3, 28.2, 50.2, 1.78.5, 9.7, 91.3, 1.87.0, 6.6, 14.8.

The new Will West card read: 19.8, 15.9, 12.2, 27.5, 50.3, 1.77.5, 9.6, 91.3, 1.88.0, 6.6, 14.8. All the corresponding numbers were either identical or well within the permissible variance.

One of the major advantages of fingerprint identification is the ease of record-taking. Other unique features of human physiognomy have been suggested as possible bases for identification systems: the pattern of the retina, the pores of the tongue, brain waves, voice prints, and even nose prints. When fingerprints have been unavailable or impossible to obtain, forensic dentistry has proved invaluable in identifying corpses. DNA typing is now possible when a suspect has left blood, sera, or tissue at the scene of the crime. But fingerprints still provide the most practical and most certain means of identification.

In 1924 the U.S. Congress authorized the Bureau of Investigation in the Justice Department (now the Federal Bureau of Investigation) to create and maintain the Identification Division. The FBI combined the fingerprint files maintained at Leavenworth with the large collection kept by the International Association for Chiefs of Police. Thus the Identification Division began its existence with a collection of 810,188 prints.

On July 1, 1931, the federal government made it mandatory for anyone applying for one of forty thousand civil service jobs to be fingerprinted. The prints would be compared against the FBI's master file, which at the time held about two million cards. The collection grew rapidly—eleven years later, on May 24, 1935, the Identification Division filed its five millionth card.

According to FBI records, the Identification Division's ten millionth card, received on January 31, 1946, was that of child actress Margaret O'Brien, taken when the nine-year-old star was touring FBI headquarters in Washington, D.C. As of February 1, 1991, the FBI had a total of 193,137,999 prints in its files, of which 107,058,738 were in the criminal division.

Today, a century after Galton, Henry, and Vucetich, along with their imitators, adapters, combiners, improvers, simplifiers, and translators, the field of dactyloscopy is undergoing dramatic change. The computer has brought a degree of speed and accuracy to fingerprint analysis that could never be hoped for before. The FBI began testing computers in its Identification Division in 1972, and by 1980 it began computerizing the entire criminal fingerprint file. On June 5, 1989, computer processing of fingerprint cards went on line, and response time for the average request was immediately cut from two weeks to one day.

Here is a brief overview of the art and science of fingerprint classification, both historical and technical. Feel free to skip over the technical parts if you are not planning to classify any cards in the near future.

The intellectual basis for using fingerprints for identification rests upon three premises:

1. Fingerprints are unique. For one finger alone, the number of possible patterns is in the hundreds of billions. This reliance on uniqueness is supported by the empirical fact that of the millions of fingerprints on record, no two have ever been found to be the same. Even identical twins with identical DNA have different fingerprints.

2. Fingerprints do not change. The prints of a small child will be recognizably the same in old age, allowing, of course, for the growth of the finger from childhood to adulthood and for any scars that may be acquired along the way.

3. A fingerprint will fall into one of several recognizable categories. This allows fingerprints to be classified according to an invariant system and permits one fingerprint to be found among the millions on file.

The method for taking fingerprints devised by the early fingerprinters is pretty much still in use today, though it is being superseded by more technical means. We will look at these further on. In the traditional method, printer's ink is spread evenly on a glass plate with a rubber roller; the person to be fingerprinted then rolls his fingers one at a time on the plate and then onto a fingerprint card that is usually in some sort of holder. After the individual fingerprints are taken, each hand is inked as a whole and pressed onto the card. The full handprint, called a flat print, is taken as a means of checking that each individual print is indeed from the finger indicated by its position on the card.

Printer's ink or something similar is used because it dries quickly and because when applied to the finger it adheres only to the ridges and does not flow into the creases. Thus a clear impression is made. Stamp-pad inks, which have been used when nothing better is available, tend to give blurred impressions. There are also various fluids for taking prints. These leave the fingers clean and develop an image on the card, but they tend to be expensive and more difficult to use.

The standard fingerprint record card, in use throughout the world, is an eight-inch square of medium-weight cardboard. It was devised in 1908 by P. A. Flak, a fingerprint expert at the Library Bureau Company of Chicago, who designed the card to be easy to file and retrieve. The actual information provided at the top of the card varies from place to place and according to whether the card is to be used for criminal identification, a military record, or civilian use (for a job application, for instance). In any case, whatever the category or the recording agency, if the print is taken in the United States a copy will probably be sent to the FBI and filed in the appropriate national registry. And while Flak's card is still in use throughout the world, more and more police departments are turning to photographic or electronic means of taking prints.

The systems devised by Henry and Vucetich for locating one card out of the millions in the files form the basis for everything that followed. Dactyloscopy now recognizes five basic patterns within fingerprints: arches, tented arches, radial loops (slanting toward the thumb), ulnar loops (slanting away from the thumb), whorls, and, finally, accidentals—traits that do not fit within the first five categories. The FBI's classification system further divides the loop into the double loop and the central pocket loop. Experience has shown that the loop pattern is the most common, making up more than 65 percent of the whole. Whorls make up about 30 percent, and arches and tented arches the remaining 5 percent. The accidentals account for fewer than 1 percent. The actual percentages vary from country to country. The pattern of the fingerprint is formed by the raised ridges that apply the ink to the card, or that leave the grease or sweat on the window pane or the household object.

Arches have a moundlike contour, with ridges that run from one side of the print to the other. Tented arches come to a point or spike. Usually a few extra arches curve over the spike.

Whorls are circular or spiral and form a complete oval around a central point.

Loops are more strongly curved than arches, with ends that enter and leave the print from the same side of the finger. Radial loops slant toward the thumb, and ulnar loops slant away from the thumb.

Of course, like most things in nature, it is never that simple. The ridge-lines of fingerprints occur in randomly curved patterns that probably evolved in order to create a skin surface that made it easier to grasp things. Nature was hardly thinking about the uniformity of pattern design. Whereas some fingerprints clearly form whorls, loops, or arches, many resemble a whorly arch, or an arched loop. The books on fingerprint classification devote many pages to demonstrating why this is a whorl while that is certainly a loop. And in order to make the classification system work, it is necessary that every fingerprint examiner see every fingerprint in the same way as every other member of the clan.

Rules had to be drawn up for the exceptions and rarities. Because the classification system is based on all ten fingers, what do you put down if one finger is missing? (It is given the same classification as that of the corresponding finger on the other hand.) What if a finger is so badly scarred as to have no usable print? It still has some pattern that can be recorded; and, if it is that badly scarred, it may well be unusable as a finger. And what if a hand has six fingers? Which one do you eliminate?

The Henry system and its cousin, the Vucetich system, as modified over the years, are the fingerprint classification systems in use over most of the world. To use them requires training in a process

Arch

Tented Arch

Whorl

Accidental

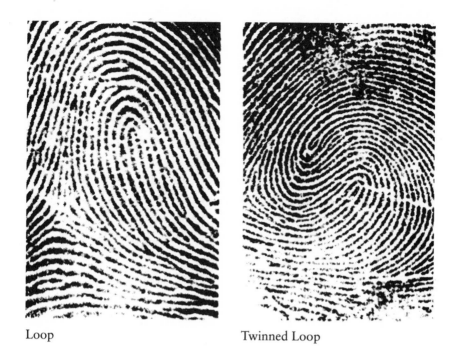

Loop   Twinned Loop

that is neither self-evident nor simple. As Peter Laurie says in his book *Scotland Yard*, "To the layman, Henry's system appears to be one of the most obscure inventions of the human mind." Still, any two people properly trained in the system will come up with the same classification for the same fingerprint card.

The basics of the system are easy to understand. What makes them forbidding to the layman is their somewhat arcane logic and their resemblance to mathematics.

In the Henry system, primary classification is based on the number and location of whorls among the prints. The fingers are numbered consecutively from the right thumb to the left little finger, and a numeric value given to each whorl, as follows:

RIGHT HAND

1/thumb	2/index	3/middle	4/ring	5/little
16	16	8	8	4

LEFT HAND

6/thumb	7/index	8/middle	9/ring	10/little
4	2	2	1	1

If the pattern on any finger is not a whorl, its value is 0. Now comes the fun part. First we make a fraction by adding the value of every print: even-numbered fingers are added together to create the numerator (the top number of the fraction), and odd-numbered fingers are added together to make up the denominator (the bottom number). A fingerprint card on which none of the fingers is a whorl would be classified 0/0. But that does not look right somehow, and it was decided that a 1 be added to both the upper and lower numbers of the fraction. That way the smallest fraction obtainable became 1/1 and the largest became 32/32. Note that this is not really a mathematical process—mathematically 32/32 is identical to 1/1 because both equal 1. In this case, if you simplify your fractions, you'll lose your data.

This process produces a total of 1,024 possible fractions that divide the collecting agency's indigestible collection of fingerprint cards into 1,024 separate files. If the collection were made up of, say, a half-million cards, each file would hold about 500 cards. This number would take a while to go through but would not be altogether unmanageable. On the other hand, when the collections grow to millions of cards, further winnowing classifications are needed.

Fingerprints can now be taken electronically or optically. These methods, combined with rapid computer recognition, mean that your finger can be used for instant identification at ATM machines or for entry to secure areas. And to further reassure those with a macabre sense of insecurity, the machines can distinguish between a living finger and a dead one. So cutting off someone's finger will not enable criminals to access their ATM accounts.

AFIS is the generic name for any automated fingerprint identification system. Today all AFIS systems are computer-generated,
using sometimes different logarithms to achieve the same results.
While several states have set up their own AFIS systems, the U.S.
Integrated Automated Fingerprint Identification System, run by
the FBI, holds all the fingerprint sets collected in the United States.
AFIS systems are very useful when two fingerprint cards are being
matched, as when someone applies for a job and her fingerprints
are submitted for a security check. If she has ever been printed
before, a match between the two cards will almost certainly be
found. By instantly narrowing the selection down to about fifty,
the AFIS system can save a lot of time in the search for a single
print. Still, when seeking a match for an imperfect latent print
lifted from a crime scene, the expertise and judgment of a trained
fingerprint examiner are essential.

6 : The Fickle Finger

> . . . It is an ever-fixed mark,
> That looks on tempests and is never shaken.
> —William Shakespeare

IN ALLAHABAD, INDIA, sometime in the 1880s, a Kayasth (member of the Kayastha caste) left a bloody thumbprint on a brass water jug found near the body of a man he was suspected of murdering. His alibi—that he had been elsewhere with two companions—was disposed of by the judge with these words: "In this land of lies, an ounce of good circumstance is worth many pounds of oral evidence, and even if, instead of two, two hundred Kayasths swore that they had sat in a circle around the accused from 6 P.M. to 6 A.M., it would be nothing in my mind compared with the unexplained bloody thumbprint."

Assuming that the thumbprint was indeed the defendant's, the judge had a point.

On May 5, 1905, twenty-three-year-old Alfred Stratton and his twenty-year-old brother Albert were brought to trial at London's Old Bailey for the vicious bludgeoning murders of Thomas and Ann Farrow, an elderly couple who managed Chapman's Oil and Colour Shop on High Street in the London borough of Deptford. The prosecution accused the brothers of entering the shop early on the morning of Monday, March 27, that year. There, it was charged, they had bludgeoned Thomas to death and had gone upstairs to the Farrow flat where they then attacked Ann. Perhaps they were searching for concealed money—the Farrows were reputed to possess a secret stash. If so, the brothers were disappointed—they found only a small cash box, which they emptied and tossed under the bed.

At 7:15 that morning, Henry Jennings, a milkman, and his "boy," Edward Russell, saw the two men leaving the shop. Russell yelled after them that they had left the door open, but they did not stop. Although mortally wounded, Thomas Farrow must still have been alive then, because at 8:30, when the shop assistant arrived, the shop door was closed. Farrow may have closed it for fear that his assailants would return.

The assistant went immediately to the nearby house of Mr. Chapman, the store's owner, and together they went to the back of the shop, pushed open a window, and climbed in. They found the body of Thomas on the floor in front of them. He had horrible wounds to his head. Ann was found unconscious and severely beaten in the upstairs bedroom; she died four days later without regaining consciousness. The cash box was found under the bed; near it lay a sixpence and a penny.

The front door had not been forced, indicating that Farrow's assailants had rung the bell, probably at around 7 A.M., and that he had come downstairs to let them in. The shop did not open until 8:30, but painters on their way to work often came by early to pick up supplies. So Farrow would have had no reason to be alarmed by an early-morning wake-up.

The culprits, Scotland Yard decided, were local men. The suspects knew that Farrow often answered his door an hour before the store opened. They shared the popular belief that the Farrows kept a stash of money upstairs, and they may have been motivated to kill for fear of identification. But which local men were they?

A young lady, Miss Ethel Stanton, saw two men running down Deptford High Street at about 7:15 that morning. She averred that one of them was a man she knew as Alfred Stratton. She did not get a good look at the other, but she noticed that he wore a brown overcoat.

The cash box was brought to Detective Inspector Charles Collins of the recently formed Fingerprint Bureau of Scotland Yard. He examined it and found a clear thumbprint on the inner tray. In those days the men of the Bureau spent most of their time comparing the ten-cards (as the cards holding the carefully taken prints of all ten fingers were called) of newly arrested suspects against the ten-cards already in their collection. Actually apprehending a murderer through a thumbprint left at the scene would open a whole new world of possibilities for the fingerprint men.

So-called elimination prints, taken in order to rule individuals out, were taken from Chapman and his assistant, the victims, and a policeman who may have touched the box. Because the print belonged to none of them, it was reasonable to assume that it belonged to one of the killers.

The Stratton brothers, Alfred and Albert, were what the police referred to among themselves as "a pair of young villains." And because they had taken off for parts unknown shortly after the crime, interest immediately focused on them. They were quickly picked up and put into what the British call an "identification parade." Still, the milkman and his boy were unable to pick them out. Ethel Stanton, however, quickly identified Alfred. So the scales appeared to be evenly weighted—Miss Stanton on one side, and the milkman and his boy on the other.

When Inspector Collins compared the Strattons' fingerprints with the one found on the cash box, he found a match for Alfred's right thumb. So Scotland Yard knew it had the right men, but would that be enough for a conviction?

Fingerprint evidence had never before been tested in a murder case. The only previous appearance of fingerprint evidence in a British courtroom had been in a September 1902 burglary trial. The parlor maid had found the billiard room of the house known as Denmark Hill broken into. As she testified: "I also observed a thumb mark on the sash of one of the windows that had been painted—it was not there the night before, and I called the attention of the police to it when they came." Detective Inspector (then Sergeant) Collins examined and lifted the print. Some months later, when a habitual burglar named Harry Jackson was arrested and fingerprinted, Collins concluded that his was the thumb that had left the billiard-room print. Collins's testimony at the trial and his explanation of fingerprint evidence had been enough to convict Jackson. In the few other cases in which fingerprint evidence might have figured, the suspects had confessed before trial.

Richard Muir was given the task of prosecuting the Stratton murder case. He would call Ethel Stanton, who saw Alfred Stratton running away from the scene of the crime. Mr. Rooth, for the defense, would question the milkman and his boy, who had a close look at the perpetrators but couldn't identify either Stratton. Stalemate. It would take the evidence of the thumbprint to win the case.

Because this would be a test trial for the acceptance of fingerprint evidence in capital cases, Muir had to be certain that both the judge and the jury understood and accepted it. He met with Detective Inspector Collins to learn more about fingerprints and how best to present them in court. They made photographic enlargements of both the crime-scene print and Alfred's right thumbprint.

Muir saved his fingerprint evidence for last. Collins testified that he found eleven matching characteristics between the print on the cash box and Alfred's print; the defense pointed out the differences in size between the two prints. Collins replied that these differences were accounted for by variations in pressure. To illustrate his point he took several impressions of a juror's thumb and showed the effects of varying pressures on the resulting prints.

The defense had a rebuttal. Dr. John Garson, vice president of the Anthropological Institute, a defender of Bertillonage and a self-appointed "expert" on fingerprints, spoke of his medical practice and of his experience measuring corpses at the Institute. He declared that

> it is very obvious on looking at the prints that in the upper print it [an area he called the "lake"] is much longer and proportionately narrower than in the lower print. I have measured the lengths and I find that the length of the lake in the upper print is thirteen millimeters, while in the lower it is eleven millimeters—that is a difference of eleven-point-eight percent.

Under cross-examination, when reminded that anthropological measurements taken by different persons seldom agreed precisely, Garson took it as a personal attack. He responded:

> I agree to a certain extent that two persons measuring the same part of the same body get different results. Very fine measurements were required, only a difference of a millimeter being allowed in the principal measurements. When the system of measurement was abandoned in October, 1901, my services were dispensed with and I have not had any connection with the Finger Print Department since that date. . . . I saw that Inspector Collins had spoken, according to the papers, a great deal of nonsense about

finger prints . . . about the number of cases you would require before you found the same finger print occurring twice, a mathematical calculation of the chances. . . . I, as a scientific man, came to the conclusion that it was nonsense.

A powerful indictment, but it did not have the desired effect. Once the prosecutor pointed out that Garson had written to both the defense and the prosecution offering his services, his "scientific" conclusion that fingerprinting was nonsense was less convincing. "I am an independent witness," Garson insisted.

"An absolutely untrustworthy one, I should think, after writing two such letters," was Justice Channell's damning reply. Nonetheless the judge cautioned the jury against relying on the fingerprint evidence alone and suggested it look over all the evidence, both for and against the defendants. The jury was out for two hours before returning with a guilty verdict.

The Deptford murder trial eased the way toward courtroom acceptance of fingerprint evidence. By 1909, New Scotland Yard was able to say that there had been many cases in which "finger print evidence has been given in court, admitted by the judges, and accepted by the jury."

The ability to associate newly apprehended felons with their prior criminal record was a great step forward. But fingerprint identification offered more than that. As Henry Faulds had demonstrated as early as 1879, it was possible to catch criminals by identifying the fingerprints they left behind at the scene of the crime.

While the Henry and Vucetich classification systems were useful for identifying felons who had already been apprehended, the process of matching single prints lifted from a crime scene was still a time-consuming task. If other evidence were available (an eyewitness identification or a known *modus operandi* of the criminal, for instance) and the search quickly narrowed, fingerprints might serve to positively identify the suspected felon.

(Latin for "the manner of operation," the *modus operandi* file lists criminals by their usual method of committing their favorite crimes, such as forcing entrance by a cellar door with the aid of a chisel. A criminal who once finds that a given method works will tend to stick to it, and thus provide the police with one more handle for identifying him.)

For a while the courts were cautious about accepting fingerprint evidence that was uncorroborated by other evidence or testimony. But as the novelty wore off, fingerprints were seen as being even more reliable than eyewitness testimony.

At the turn of the century, dactylography experienced a vogue in Europe. Books and pamphlets on the subject, both popular and professional, appeared in most European languages. By 1910 fingerprint evidence had been instrumental in solving cases in France, Germany, Belgium, the Netherlands, Czechoslovakia, Switzerland, and Austria-Hungary. In 1905 Austria-Hungary became the first continental country to adopt the Henry system. Colonel Alfred Redl, head of the Kundschafts Stelle, the military counterintelligence service, kept highly polished, silver cigarette cases and candy dishes in his office in order to entice his visitors and thus unobtrusively obtain their fingerprints. In 1913, when it was revealed that Redl had for years been an agent of Russian military intelligence, he committed suicide.

At about 2 P.M. on the morning of September 19, 1910, Clarence B. Hiller of West 104th Street, Chicago, was awakened by his wife in the bedroom they shared with the two youngest of their four children. When she nudged him awake to tell him that the gaslight in the hall was out, he rose to investigate. If the gas were on but the flame out, it would be a serious hazard.

When Hiller reached the landing, he found it already occupied. He and an intruder tussled until both of them fell down the stairs. The intruder rose first, pulled a revolver, and fired two shots at Hiller, mortally wounding him.

On a street about a mile away, four policemen, their shift over, sat on a bench awaiting a late-night streetcar. They heard nothing of the commotion at the Hiller house, but they did see a man scurrying along the street. Being policemen, they asked him what he was doing out at that hour. When his answer proved unsatisfactory, they searched him and found a loaded gun in his pocket and bloodstains on his clothing. He claimed to have cut his finger when getting off a streetcar the day before. The cops took him back to the station house on the streetcar.

The man's name was Thomas Jennings. He seemed unable to tell a consistent story and wasn't even sure of his address. First it was 1244 State Street; within the hour it became 577 Twelfth Street.

The next day, when the police investigated the murder of Clarence Hiller, they found that the killer had entered by way of the kitchen window. He had removed the screen and boosted himself in by way of the porch railing. Once inside he had crept upstairs, attempted to molest the oldest daughter in her bedroom, then retreated to the hall where he encountered Hiller.

On the porch railing—the freshly painted porch railing—just at the place on the rail that someone would grasp in order to hoist himself through the kitchen window, the police found the perfect imprint of the four fingers of a left hand. The fingerprint experts compared those prints to those of Thomas Jennings and declared them a match.

There was a great deal made of the fingerprints at Jennings's trial. Fingerprint evidence was new, and both the judge and the jury had to be convinced. Blown-up photographs of the prints were introduced into evidence, and four qualified experts, two of them trained at Scotland Yard, testified they were each positive that the fingerprints on the porch railing were those of the suspect. Jennings was convicted and, on February 1, 1911, sentenced to death.

The sentence was promptly appealed. The main argument of the appeal was that the fingerprint evidence had been improperly introduced. "It is earnestly insisted," the defense maintained, "that this class of testimony is not admissible under the common-law rules of evidence, and as there is no statute in this State authorizing it, the court should have refused to permit its introduction."

But the appeals court was unconvinced, ruling among other things that: "Expert evidence is admissible when the witnesses offered as experts have peculiar knowledge or experience not common to the world, which renders their opinions, founded on such knowledge or experience, an aid to the court or jury in determining the questions at issue."

And so the sentence of death was upheld, and Thomas Jennings became the first person in the United States whose conviction for a capital crime was based on fingerprint evidence. On February 16, 1912, he was hung by the neck until dead.

When a person touches an object he often leaves behind a fingerprint consisting of sweat, body oil, grease, dust, blood, or some other material. Inadvertently left fingerprints are of three types—visible, plastic, or latent.

Visible prints are the sort that small children leave on refrigerator doors when they retrieve the milk after eating peanut butter and jelly sandwiches. The transmitting medium may also be grease, dirt, ink, paint, or blood.

Plastic prints are indentations left in objects that are soft or pliable enough to take and hold an impression. This might be warm wax, soft soap, butter, putty, or drying paint.

Latent prints are left by perspiration or grease and tend to be invisible. But if light strikes a glossy surface at just the right angle, such a print may be seen.

The fingerprint technician's job is to locate fingerprints left at a crime scene, develop any latent prints, and preserve and prepare all useful prints for observation and identification. For elimination purposes, he will begin by taking the fingerprints of everyone with a legitimate reason to be at the scene, including any police officers. He will then search for latent prints, "lift" the prints he finds onto index cards when possible, and mark each card with the appropriate case number, time, and location.

The earliest method for finding and developing latent prints involved the use of a fine powder such as carbon black, which was dusted onto the area to be examined with a soft brush or an insufflator (much like a perfume atomizer). If a latent print were present, the powder adhered to it when the excess was blown away. Then the print could be lifted with transparent tape and placed on the index card.

Fingerprints may be left on almost anything. And these days they can be lifted or otherwise secured from just about any surface on which they are found. The length of time a latent print will last varies in duration from moments to centuries. On a nonporous surface in cold, dry weather, the life of a latent print will probably be brief. On porous surfaces in damp weather, a print may last for weeks. On newsprint or other nonglossy papers, it may be there for the life of the paper.

The choice of powder to be used depends mainly on the material and the color of the object to be examined. The powder should be one that sticks to fingerprint moisture or oil but not to the object itself, and that shows up well in contrast to the background color. Usually a black powder is used on light-colored surfaces and a grey powder on dark or black surfaces.

But there are many powders available, and selecting the best one for a specific surface is, even today, more of an art than a science. For a time, an organic powder called "dragon's blood" was very popular because it was one of the few powders that worked at all on paper. But it has fallen out of favor.

Fuming a print with an iodine vapor is a method devised in order to develop fingerprints on paper or unfinished wood. Iodine vapor turns paper a light brown color and turns fingerprints a much darker color. Iodine must be handled with care, as the vapors are both corrosive and poisonous. Prints developed this way must be either "fixed" with another solution (7,8-benzoflavone works well) or photographed almost immediately—as the iodine evaporates, the print fades away.

Another method of developing prints on porous surfaces such as paper or fabric depends on the salts—among them sodium chloride (ordinary table salt)—left behind when sweat dries. When the area being tested is treated with a solution of silver nitrate, it reacts with sodium chloride to produce sodium nitrate and silver chloride. The silver chloride is light sensitive. When exposed to sunlight or any other source of ultraviolet radiation, it reduces to metallic silver and the print then appears in a color that is somewhere between brown and black. As with iodine fuming, the print must be photographed as quickly as possible; the background will continue to blacken, and the contrast will be lost after a short while.

Silver nitrate has now been replaced by a "stabilized physical developer," or SPD, a silver-based solution that works well on surfaces that are, or have been, wet. The prints appear in shades of grey, and, just as with silver nitrate, must be photographed as soon as possible.

Ninhydrin is a chemical that turns purple in the presence of, among other things, the amino acids present in latent fingerprints. This was first noticed in 1954 by two Swedish chemists, Oden and von Hofsten. Ninhydrin is particularly effective for developing latent prints on paper or other porous materials such as cardboard, wallboard, or untreated wood. If the material has been kept under dry conditions, it can pick up prints after a considerable length of time has passed. Prints that are more than thirty years old have been discovered and developed using ninhydrin.

In May 1977, Fuseo Matsumura, a hair and fiber expert at the Saga Prefectural Crime Laboratory of the National Police Agency of Japan, was mounting hairs—the evidence in a criminal case—on a glass slide when he noticed that his own fingerprints were developing on the slide. A little investigation showed that fumes from the cyanoacrylate glue he was using had brought out the fingerprints. He passed this information on to a co-worker, a latent-print examiner named Masato Soba.

In May 1979 a British fingerprint examiner, L. W. Wood, of the Northampton Police Department, noticed his fingerprints on a film tank he was repairing with superglue. The British police establishment quickly set about developing this new technique.

In September 1979, Paul Norkus and Ed German, fingerprint examiners with the U.S. Army Crime Lab, saw a demonstration of superglue fuming at the National Police Agency of Japan. They experimented with it and found it could raise prints from surfaces that previously had proved difficult or impossible. Prints raised with superglue may be seen by dusting them with powder or by washing them with a laser-sensitive dye that allows them to be viewed and photographed in laser light.

In the summer of 1985 a terror was visited upon the city of Los Angeles. A killer, dubbed the "Night Stalker" by the press, was attacking his victims indiscriminately. Without regard to race, status, or age, he attacked single women, couples, the young, and the old. These attacks usually occurred in the victims' homes but on occasion took place in their cars. The men were beaten to death; the women were sexually molested and then stabbed. The female victims ranged in age from six to over eighty. Most of the women died, but a few survived to give a description of the attacker: he was young, Hispanic, tall, thin, and had an angular face. By August 22 the Los Angeles police credited him with at least fourteen killings and sixty-eight other felonies, including multiple counts of rape and assault.

On August 24 the Night Stalker traveled outside his usual stalking area and broke into the home of Bill Carns and his fiancée Inez Erickson in Mission Viejo, fifty miles south of Los Angeles. He shot Carns, raped Erickson, and fled in a stolen orange Toyota. A local teenager, seeing the car cruising the neighborhood, took down its license number. The car was found by the police two days later in the Rampart area of Los Angeles.

The forensic team processing the car decided to try the still-new fuming technique to see if they could raise any latent fingerprints. They put a saucer of superglue in the car, closed the windows, and left it overnight. The next day they used a laser to search for prints and found one single usable fingerprint. A copy was sent to the new state computer system in Sacramento, which came up with a match almost instantly: a twenty-five-year-old car thief named Ricardo "Richard" Leyva Múñoz Ramírez.

On August 30 copies of Ramírez's picture were broadcast by local television stations and spread around Los Angeles. On August 31 he was spotted after leaving the Greyhound bus station in East Los Angeles. He escaped his original pursuers but was caught a couple of miles away while trying to steal a car. By the time the police arrived they had to rescue him from an angry mob.

On November 7, 1989, after a trial lasting more than a year, Richard Ramírez, the Night Stalker, was convicted of thirteen counts of murder, five of attempted murder, eleven sexual assaults, and fourteen burglaries. He was sentenced to death in San Quentin's gas chamber. As this is written, he awaits execution, though the method of implementation has been changed to lethal injection.

The apprehension of the Night Stalker was a result of two modern forensic techniques, new at the time: fuming with super-glue and the computerization of fingerprint files.

In 2008 the British scientist John Bond revealed a new technique for developing latent prints. It works on metal surfaces, where it detects minute bits of corrosion caused by the salt in fingerprint deposits. It is especially effective for use on shell casings. While the technique is simple, it is not obvious. The bit of metal to be tested is charged to an electrical potential of 2,500 volts. Then a similarly charged powder comprised of fine ceramic beads that are coated with an even finer black powder is poured over it. The black powder adheres to the points where the salt corrosion is present. When the metal is heated, the powder is baked on. *Voilà!* The print is revealed and may be photographed.

Over the years fingerprint examiners have developed methods for handling specific materials to be processed for latent prints. Different plastics, for example, are treated in different ways. Lately a new problem has surfaced: recycling. When recycled plastics are melded together to be turned into new products, the resulting amalgam is a unique blend with no predetermined treatment. And each batch is different from the last. Thus are the lives of fingerprint experts saved from the ennui of mundane existence.

After a century of supposed reliability, the identification of la-tent prints has recently come into question. The problem is not with the uniqueness of fingerprints; so far no one has found two identical prints that were not from the same finger. But the skills of some fingerprint examiners are not up to the task of correctly identifying the twisted and smudged partial prints that are often the only ones found at a crime scene. Further, some fingerprint examiners seem unable to recognize their limitations.

In January 1996 the body of fifty-one-year-old Marion Ross was found in her cottage in Kilmarnock, Scotland. She had multiple stab wounds to her eyes and throat. A fingerprint examiner from the Scottish Criminal Record Office identified a print found on a gift tag at the scene as being that of David Asbury. He was arrested and placed on trial for the murder. In addition, one of the other prints found, this one on a door frame in the house, was identified as that of Constable Shirley McKie of the Strathclyde Police Department, one of the investigating officers.

When Asbury's house was searched, a print of the victim, Marion Ross, was found on a biscuit tin in his bedroom. The tin contained £1,800. Asbury insisted that the money was his, and no one could say that it was not. But what about the print on the biscuit tin? Because it was possible that Asbury's print could have been left on the tag when he worked in the Ross house some time earlier, the print on the biscuit tin—Marion Ross's—became a key piece of evidence against him.

But McKie's print on the door frame presented a problem. She was not supposed to enter the house, and she certainly was not supposed to enter the house without wearing latex gloves. When questioned about this, she had a simple answer: in fact she had not entered the house. How could she be believed? As Ian MacWhirter put it in an article in the *Glasgow Sunday Herald* of February 26, 2006:

> Finding an unknown fingerprint at the scene, they sought to identify it so as to eliminate its owner from the inquiry.
>
> The Scottish Criminal Record Office checked local police officers who might have been there and found a similarity with that of Detective Shirley McKie.
>
> Hey, get Shirley to agree that she had made an unauthorized visit to the scene, and bingo. Result!

But McKie wasn't having it. Difficult.

Woman in a man's job; had to fight to win professional recognition, and so on.

Even her dad, a former police superintendent, didn't believe her at first. Admit it, he said, put it behind you. This made her even more determined.

At the Asbury trial, to the consternation of the prosecutor, McKie testified to what she knew to be true. Either she was lying to cover up what might have been an innocent mistake, or the fingerprint examiners were wrong, in which case they might also be wrong about the Ross print. And how many other fingerprints had they, and possibly the entire Scottish Criminal Record Office, been mistaken about over the years?

Despite doubts about Shirley McKie's testimony, the jury found David Asbury guilty. He was sentenced to life in prison, and that would seem to have been that. Except that in March 1998 McKie was arrested in her home, taken to the Ayr police station, strip-searched, and charged with perjury. If convicted, she would go to prison for eight years.

In order to prove that she had not lied, McKie would need to prove that the fingerprint found at the crime scene was not hers. Since the fingerprint indeed was *not* hers, she should be able to find someone to say so, someone with the necessary credentials to be believed. She got in touch with Patrick A. Wertheim, a highly regarded fingerprint analyst with the Arizona Department of Public Safety. Wertheim listened to McKie's story and then flew to Glasgow to look at the prints for himself. He was shocked to discover that the prints did not match—they were not even close. As he wrote sometime after McKie's perjury trial:

For an "expert" to claim an identification between the crime scene mark in this case and Shirley McKie's fingerprint is sheer incompetence. To testify knowingly to this erroneous identification, as some believed happened at Ms. McKie's trial, is perjury. For the police to cover up the mistake is despicable. For a whole police administration to blindly support their "experts" without seeking a competent outside review is foolhardy.

But in May 1999 the authorities went ahead with their trial. The Scottish Criminal Record Office stuck to its opinion while Pat Wertheim and his fellow experts, David Grieve of the Illinois State Crime Lab and Allan Bayle of New Scotland Yard, forcefully rendered their opinions that the two prints did not match. It took the jury less than an hour to find McKie not guilty.

McKie had won this battle, but the war was far from over. Officials at the SCRO were not about to take their defeat lying down. They mounted a campaign of propaganda and disinformation in which they attacked Patrick Wertheim's credentials, accused Shirley McKie of further unprovable indiscretions, and sent memos to the law enforcement community in which they touted their own "integrity." McKie was fired from the Strathclyde Police Department in December 1999 for "medical" reasons.

When McKie attempted to sue the police department for wrongful termination, the suit was dismissed because, as Lord Emslie of the Court of Sessions put it, she had failed to prove that her superiors had acted with malice.

Then the story took a new and startling turn. Pat Wertheim was asked by BBC Scotland to review all fingerprint evidence in the Ross case. In examining the court exhibit that compared Marion Ross's thumbprint with the one found on the cash box in David Asbury's bedroom—enlargements of each print mounted side by side with red lines connecting the "points of similarity"— he quickly saw that they were not similar at all. It seemed that no effort had been made to connect one end of a red line to anything that looked like the other end. It was not so much a question of a bad match as one of fraud.

Wertheim and Bayle examined the actual biscuit tin and could not find Marion Ross's prints anywhere on it. And Asbury's mother continued to maintain that the tin was her son's, and that the money in it had not been stolen from Ross but was in fact his life savings. He had planned to buy a car.

It looked as though the police, once they had settled on Asbury as their prime suspect, had gone to unusual lengths to establish his guilt. There was no forensic evidence against him save for one fingerprint. It is no wonder they did not want doubts raised about it.

Was Wertheim sure about his new finding? He was.

> . . . and my career is on the line on this. And I'm not putting my career on the line on a gamble, I don't gamble with that. I am right. My position has been confirmed by other experts from other countries all over the world. The SCRO is wrong. They have no one to support them in this. No one at all except their own internal people who are applying the whitewash to this thing.

In August 2002, David Asbury was released from prison. After spending five years locked up on bad evidence, he believed he was entitled to an apology. He never got it.

In February 2006 a case brought by Shirley McKie against the Scottish government for malicious prosecution was settled for £750,000 with no admission of liability. The settlement was reached just before the case was to come to court. As the *Glasgow Herald* put it, it was "a clear case of throw the money and run."

In Boston, on May 30, 1997, in the back yard of a house on School Street, a police officer grappled with a burglar who then grabbed his gun, shot him twice, and took a shot at someone leaning out of a second-floor window. The burglar then leaped over a fence and ran away, leaving his baseball cap behind. He forcibly entered a nearby home and held a family hostage for a couple of hours while hiding from his pursuers. He then fled into the night, leaving behind his gun and a white sweatshirt.

Two weeks later the police officer looked at a photo lineup of eight pictures and identified twenty-seven-year-old Stephan Cowans, a known burglar, as his assailant. On July 2 both the officer and the witness from the second-floor window picked out Cowans from a live lineup. Curiously, the family that had been held hostage, and who had the best and longest look at the burglar, failed to identify Cowans as their assailant.

Fingerprint experts from the Boston Police Department examined a mug from which the burglar had drunk while at the hostages' home. They found a latent print on the rim and identified it as Cowans's left thumbprint. Cowans was easily convicted on the combination of eyewitness and fingerprint evidence and was sentenced to thirty-five to fifty years in prison.

From his prison cell, Stephan Cowans continually maintained his innocence and managed to interest the Innocence Project in his case. Founded in 1992 by attorneys Barry C. Scheck and Peter J. Neufeld at the Benjamin N. Cardozo School of Law in New York City, the Innocence Project assists prisoners who might be proven innocent through the new science of DNA testing. As of 2008, 232 people in the United States have been exonerated by DNA testing, including 17 who served time on death row. These people served an average of 12 years in prison before their exoneration and release.

The Innocence Project arranged for DNA testing of the mug and the baseball cap. And they found that where the DNA on each matched the other, neither matched that of Cowans. The Suffolk County district attorney insisted that the white sweatshirt also be tested. It was, and again the DNA on it matched the mug and the cap, but not Cowans's. Finally, the DA had the damning fingerprint reanalyzed and found that the analysts of the Boston police were mistaken—the fingerprint did not belong to Cowans.

On January 23, 2004, after six and a half years in prison, Stephan Cowans returned home. Without the fingerprint evidence, Cowans almost certainly would not have been convicted—he might not even have been brought to trial. On October 26, 2007, Stephan Cowans was shot to death in his home. The reason is unknown, and no one has been charged with the crime.

On March 11, 2004, the commuter rail system serving Madrid, Spain, came under coordinated attack by terrorists. Ten explosions occurred on four trains at the height of the morning rush hour. One hundred ninety-one people were killed in the blast, and 1,460 were injured. Shortly afterward the Madrid police searched a stolen van parked near the Acala de Heres train station, the point of origin of all the trains involved in the bombing. In the van they discovered a blue plastic bag holding bomb detonators. Fingerprints were lifted from the bag, and digital images of several of them were sent to the FBI, which then ran them through AFIS and came up with 20 possible matches. From these 20 a senior fingerprint examiner named Terry Green picked out the prints of Brandon Mayfield, a lawyer from Portland, Oregon, as an actual match. It was a question of "100 percent identification," Green said. Three other FBI print examiners agreed with him. Mayfield's prints were in the database from his time as a lieutenant in the U.S. Army.

Under the terms of the U.S. Patriot Act, Mayfield was placed under surveillance. On May 6, 2004, FBI agents arrested him on a material witness warrant in connection with the Madrid bombing. Federal prosecutors had filed a nine-page affidavit giving cause why the arrest warrant should be issued. In addition to the fingerprint evidence, the affidavit alleged that:

—Mayfield had performed legal work for Jeffrey Battle, one of the "Portland Seven," a group of homegrown terrorists who supposedly had tried to join the Taliban and had been tried and convicted of conspiracy to levy war against the United States.

—On the first anniversary of the September 11 attacks, someone in Mayfield's house made a phone call to an Islamic charity run by a Muslim man then on a federal terrorism watch list.

—Mayfield's law practice was advertised in a directory published by a man who had once been a business associate of Osama bin Laden's personal secretary.

—Mayfield attended a mosque near his home. (He was married to an Egyptian and was a convert to Islam.)

Two federal public defenders, Christopher Schatz and Steven Wax, were appointed to defend Mayfield. They went immediately to U.S. District Court Judge Robert Jones, the judge who had authorized the arrest warrant, and asked that Mayfield be allowed to go home and remain under house arrest. When Judge Jones asked the prosecutor to cite the evidence against Mayfield, the prosecutor grudgingly allowed that it was limited to a single fingerprint identification.

Meanwhile the Spanish police, who disagreed with the FBI's identification of the fingerprint, had continued to work the case and grew convinced that the FBI identification was mistaken. The United States prosecutor failed to inform Judge Jones of the conflict in identification and refused to allow the defense to have its own expert evaluate the print. He cited "national security" concerns.

Although Judge Jones agreed that even in his own courtroom fingerprint evidence had often been sufficient to send people to prison for life, he thought that someone else should at least look at the alleged fingerprint. The prosecution finally agreed to allow a fingerprint expert, one to be selected by the judge, to perform an independent evaluation. Jones chose Kenneth R. Moses, a former officer with the San Francisco police. Moses' company, Forensic Identification Services, claimed expertise in the computer enhancement of fingerprints.

On May 9, 2004, fifty-nine days after the bombing and three days after Mayfield's arrest, Moses called Judge Jones to announce his results. Although making the determination had been difficult because of "blurring and some blotting out," the San Francisco expert said, there was "100 percent certainty" that the print he had examined belonged to Brandon Mayfield.

Before Judge Jones had a chance to pass this devastating news on to Mayfield and his attorneys, the prosecutors stepped forward with some news of their own. The Spanish police had forwarded information to them that "cast some doubt on the identification" of Mayfield. The information was classified and should not be discussed in open court, they argued.

Jones invited the prosecutors back to his chambers, where they remained for twenty minutes. When they emerged they sheepishly informed Mayfield and his attorneys that Ouhnane Daoud, an Algerian national living in Spain, had been arrested by the Spanish police. They had established that the print found on the blue bag matched Daoud's right middle finger. Judge Jones ordered that Mayfield be freed.

In this and other cases described above, the problem was not one of the uniqueness of fingerprints. It is clear by now that if each individual fingerprint is not unique, it is close to it. The problems are rather ones of interpretation—fingerprints are read by human beings. Even now, when computers can do 90 percent of the selecting, the final 10 percent is still performed by trained analysts using their experience and judgment. But human beings are fallible and susceptible to being influenced by what they expect to see and are expected to see. Can there be any doubt that the FBI's misreading of Mayfield's fingerprint was influenced at least in part by the fact that he was a Muslim?

In the book *Fingerprints Can Be Forged* (1924), Albert Wehde and John Beffel took a swipe at the supposed invulnerability of fingerprint evidence. They pointed out that there were no statistics newer than Galton's and that the experts had little incentive to provide examples of problems with fingerprint evidence. Indeed, the examiners' self-interest dictated that they defend the infallibility of fingerprint experts against all attacks. And the idea of fraud in the ranks of the fingerprint takers was not to be discussed or even considered.

Wehde and Beffel were instantly and fiercely attacked by the forensic community, castigated as radical troublemakers. Beffel, who had merely written the introduction to the book, was essentially ignored, and Wehde was labeled a troublemaking jailbird (he had served a year in Leavenworth Penitentiary as a war resister before being pardoned in 1922 by President Harding). It took seven decades for the problems that Wehde and Beffel had raised to get a hearing in the courtroom.

In 1993 the Supreme Court decided a civil case, *Daubert v. Merrell Dow Pharmaceuticals*, ruling that scientific evidence presented in a court case would be admitted based on "an independent judicial assessment of reliability." The previous standard had been its "general acceptance" by the scientific community.

The court drew up a list of questions that the trial judge should consider in assessing such scientific evidence:

1. Has the scientific theory or technique been empirically tested?

2. Has the scientific theory or technique been subjected to peer review and publication?

3. What is known of the potential rate of error of the theory or technique?

4. Are there standards for controlling the use of the scientific technique, and are these standards maintained?

5. Is there general acceptance of the technique by the scientific community?

Since then the so-called Daubert hearings have been used with some success to attack several pseudoscientific corners of forensic investigation such as bite-mark analysis and handwriting comparison. But no one anticipated that it would be used to successfully attack fingerprint analysis and identification.

Following Daubert, defense attorneys made several feints at overturning faulty fingerprint identifications, pointing out that people had been sent to prison on the unsupported word of a fingerprint examiner alone. As attorneys Peter Neufeld and Barry Scheck explained in an opinion piece in the March 9, 2002, *New York Times*:

> In 1993, when the Supreme Court demanded real scientific standards for expert evidence in federal courts, some critics correctly anticipated that several criminal identification techniques would be attacked in the courts with some success: microscopic hair comparison, bite mark analysis, handwriting comparison. Few, if any, predicted what is happening now: The bedrock forensic identifier of the 20th century, fingerprinting, has started to wobble. . . .
>
> Fingerprint experts had conceded that the process they use—matching large, evenly pressured prints taken from suspects at the police station to smaller, unevenly pressured prints from crime scenes—is ultimately subjective and bedeviled by inconsistent standards. The French, for example, require that two fingerprints match at 16 points before they can be accepted as coming from the same person; the Australians, 12; and the Swedes, 7. The F.B.I. refuses to state a number at all, relying instead on case-by-case judgments.

In a proficiency test given to 156 fingerprint examiners in 1995, 56 percent of them failed to correctly identify at least one of the five fingerprints they were given. By 1998 the number was down to 42 percent. This is not acceptable for a technique that is presented in courtroom testimony as infallible.

In January 2002, Judge Louis H. Pollak took note of these findings in a murder trial at a federal court in Philadelphia, finding that there was no persuasive proof that fingerprint evidence has been scientifically validated through objective, controlled experiments. At first barring the use of such evidence, he later partially retracted his ruling, agreeing to allow fingerprint evidence to be used, but with a caution. Still, the skeleton is out of the closet—it will now be more difficult for judges to dismiss summarily a defense attorney's challenge to fingerprint evidence.

In addition to the misidentification of fingerprints, whether from ineptitude or malice, problems of fingerprint forgery and fabrication arise. Fingerprint forgery is the creation of the pseudo-fingerprint of someone, probably with the intention of leaving it behind at a crime scene. Let's say, for example, that Andrew leaves an imprint of his finger in a pat of butter. Boris freezes the butter and uses the imprint as a mold to cast a rubber copy of Andrew's finger. Boris can then go around leaving Andrew's latent print at the scene of every safe he cracks, if Boris happens to be a safecracker. This occurs in fiction more than in real life, R. Austin Freeman's 1929 story "The Old Lag" being a fine example.

Recently a real-life illustration of the technique came to light when fingerprint expert Pat Wertheim was asked to examine prints found on the canvas and stretcher of a purported Jackson Pollock painting that had been used to authenticate the work. The prints were said to match those on a paint can in Pollock's studio, and thus could be assumed to be his.

Wertheim found that the prints on the painting did indeed match those on the paint can. They matched a bit too well. Fingerprints do not change, but the amount of pressure used to apply them, or the angle at which they touch the surface or the receptiveness of the material, do change, and these prints showed none of those differences. The fingerprints were a forgery.

Wertheim said he knew of only one other case of forgery in the history of fingerprinting. In 1947, in Sofia, Bulgaria, a fortune-teller named Nedelkoff had a sideline as a safecracker. As a fortune-teller he would have his clients press their hands into soft clay and read their fortunes from the impressions. As a safecracker he would make a mold of the fingerprints of his clients and leave prints around the crime scene to mislead the police.

Fingerprint fabrication, on the other hand, is a result of police or CSI misconduct. It is the act of claiming that a print lifted at one location was actually lifted somewhere else, usually at a crime scene. This trickery is little known outside the community of fingerprint experts and is discussed only very quietly within. Wertheim gives several examples of the deed, dating from 1925 to the present. In 1928, Lloyd Fogelman was sent to prison when a fingerprint examiner showed a photograph of Fogelman's fingerprints in court and claimed they had been found on the windowsill at a burglary. It was subsequently proven that the examiner had merely photographed a fingerprint card holding Fogelman's prints and cropped out the lines and words on the form. Fogelman's conviction was overturned, but the examiner was never penalized for his indiscretion. From then through the 1950s, according to Wertheim, the FBI discovered at least fifteen more cases of fingerprint fabrication in the United States.

One of the most famous murder cases of the twentieth century involved fabricated fingerprints. The victim was Sir Harry Oakes, at the time one of the richest men in the world. This complex case has been the subject of several books but has never officially been solved.

The crime took place in Nassau, capital of the Bahama Islands, in the second week of June 1943. At around seven in the morning Sir Harry was found dead in one of the twenty bedrooms of his estate, killed by blows to the head by some object, never found, that left half-inch wide triangular depressions. While Oakes was still alive, his bed had been set on fire, and his head and genitals severely burned. The perpetrator had waited around to put the fire out before it spread, and had then sprinkled the smoldering bed and body with feathers from the pillows.

The royal governor of the islands was Edward, Duke of Windsor, the former King Edward VIII of England who had given up his throne to marry a commoner and had been given the Bahamas as a consolation prize. On being told of his good friend's death, the duke for some unknown reason telephoned Captain Edward Melchen, chief of the Homicide Department of the Miami police. Edward had apparently been pleased with the way Melchen had once handled a security detail when the duke had passed through Miami. Still, the Bahamas were British territory, and they did have their own police force. The duke asked Melchen if he could come over to assist, as "a very prominent citizen who is one of my close friends has committed suicide."

Captain Melchen and Captain James Barker, head of the Miami Police Identification Department, took the next plane to Nassau. One look at the body, one look around the room, told them that if there was one thing Sir Harry's death wasn't, it was suicide. They carefully examined the arms and hair of Harold Christie, a business acquaintance of Oakes's and the only other person who admitted to being in the house that night, and found not even a suggestion of a singe. Christie told them that after a party the evening before—ironically a farewell party, as Oakes was scheduled to leave for South America—which the Duke and

Duchess of Windsor and a few other close friends had attended, the servants had retired to their nearby cabins. Oakes and Christie had talked business for a while and had then gone to their respective bedrooms.

When the detectives made a list of logical suspects, one name stood out: Count Marie Alfred Fouquereaux de Marigny, or "Freddie," as his wife Nancy, the eldest daughter of Sir Harry Oakes, called him. Count de Marigny and his father-in-law had not gotten along. Interviewing the count, the detectives discovered that, though he had a fairly tight alibi, the hair on his arm was singed—and it looked like the tip of his beard had also been visited by fire. His explanation, that he had burned himself lighting torches to illuminate an *al fresco* dinner for a small group of friends, didn't satisfy them, especially when he added that he had changed his mind and served the dinner indoors.

The Miami detectives and the Bahama police settled on Count de Marigny as the killer and began building a case against him. Nancy de Marigny realized she'd better do something to defend her husband, who she couldn't believe had killed her father. She hired New York private detective Raymond Schindler who, despite being known as "the society detective," was one of the best in the business.

Schindler went to Nassau and met with the count. Looking over the evidence, he concluded that de Marigny was probably innocent. First of all, Schindler decided, the count was a lover, not a killer. Second was the murder itself. Someone—or perhaps two or three persons—had bashed in Sir Harry's head, hitting him savagely at least four times, then burned his face and genitals, and waited around to put the fire out and spread feathers over the bed and body. That denoted either extreme displeasure or possibly some sort of satanic ritual. If de Marigny had done it, it would have been for money, and he would have left the scene as quickly as possible.

But the Miami detectives had found one damning piece of evidence: a print of the Count de Marigny's little finger on a folding screen by the bed. They had lifted the print and penciled a circle around the spot where it had been found.

Schindler wrote up all the evidence, pro and con, and took a flight back to Washington, D.C., to visit an old friend, Homer S. Cummings, Franklin Roosevelt's Attorney General. Cummings had been a prosecutor in Connecticut, but his fairness was legendary. He agreed with Schindler that de Marigny was probably innocent and offered his help if needed. He pointed out that the fingerprint was the only substantial evidence the prosecution had. Schindler then recruited two fingerprint experts, Captain Maurice O'Neil, chief of the Bureau of Identification of the New Orleans Police Department, and Professor Leonard Keeler, director of the Illinois Crime Bureau. They flew back to Nassau with him and examined a copy of the fingerprint. The background of the print appeared to have little circles and seemed to have been taken from a smooth surface. Schindler photographed the area of the screen from where the Miami detectives claimed to have taken the print. It was not a smooth surface. It had no little circles.

"There's not one chance in ten million," Keeler said, "that this print came from that screen."

There had been bloody handprints on the wall of the room, according to the Nassau police first on the scene, but the day after the murder someone had washed the walls.

The Nassau authorities went ahead with the trial. Schindler's experts trumped the authorities' fingerprint evidence, and the Count de Marigny was found not guilty. It was thought that one of the two Miami detectives must have fabricated the print, but the point was never pursued. Schindler, fascinated by the turns the case had taken, wanted to continue the investigation on his own time, but the Duke of Windsor forbade access to several of the promising leads and effectively closed down the inquiry. No one knows why.

Attorney General Cummings later wrote Schindler a letter about the case, observing:

> From what I can see from the record, the police in Nassau fell into a mistake common to inexperienced officers everywhere. They found their logical suspect first and then proceeded to search for facts to fit him. From this first step it was easy to fall into a state of mind where the investigator makes himself blind to every bit of evidence except that which he can apply to the preconceived theory he has created in his own mind.

He didn't mention the state of mind where the investigator manufactures his own evidence to support his theory.

7 : With a Bullet

HISTORICALLY "ballistics" has been the study of the behavior of a projectile—a ball, bullet, arrow, dart, quarrel, artillery shell, or pie—anything that makes its way through the air from one place to another powered by an initial impetus. The word derives from the name of a weapon used by the ancient Romans, the ballista, an oversized windup crossbow they used to hurl heavy darts or even heavier stones at their enemies. The term is still used within the shooting community to denote the study of the effects of gravity and wind resistance on a projectile of a given shape, mass, and velocity, fired at a given angle.

Over the past eighty years or so the forensic definition of ballistics has expanded to include everything having to do with projectile weapons and the objects shot from them. Forensic ballistics is concerned with the characteristics of a weapon, the nature of its explosive charge, the shape and composition of the projectile, the effects of the projectile's progress through the barrel of the weapon, and the means of identifying the particular weapon from which the projectile was fired.

If a projectile weapon is used to commit a crime, there will be questions for the ballistics expert: Just what sort of weapon is this? Who made it, and how might it have gotten here? What sort of projectile does it fire? Is this the actual weapon from which the projectile was fired? From what distance was it fired? Is this ammunition consistent with the ammunition found at the suspect's house? Was the shooting accidental or deliberate?

Until the early years of the twentieth century, the police consulted a gunsmith when they needed information about a gun. But many of the questions of interest to criminal investigators are outside the experience of the average gunsmith. As Hans Gross explained in 1906:

> Even when dealing with a gunsmith who knows his trade, we find his knowledge restricted in most cases to being able to indicate the origin and the price of the weapon, the names of its different parts, and other mechanical details, which it must be confessed, will have in most cases a certain value. But he will not be able to say much regarding the use to which the instrument may be put, the effects which it is capable of producing, the connection existing between the arm itself and the bullet employed, besides numerous other questions of capital importance.

Gross suggested that the criminal investigator with a firearms question might have to turn to experts in other fields—"the experienced sportsman, the medical man, the inspector of musketry, the physicist, the chemist and the microscopist."

In order to encompass this breadth of expertise in one person, another fund of knowledge had to be built up, knowledge of great interest to the police but with little application elsewhere. It wasn't until the 1930s that crime labs could identify and train people who could honestly call themselves ballistics experts.

The first known apprehension of a murderer through what we now call ballistics evidence occurred in England in 1794. The victim was Edward Culshaw of Lancashire, and the suspect was John Toms. The weapon was a muzzle-loading flintlock pistol.

To load a muzzle-loader one first poured a measured amount of gunpowder into the barrel, then rammed in a wad of paper. Next, one dropped in the ball, then rammed down another wad of paper to jam the ball in so it would not simply roll out. Then one put a dash of gunpowder in the pan, which connected to the gunpowder in the barrel through a thin tube bored through the side. To fire the thing, one cocked a hammer holding a flint and pulled the trigger. As the flint hit the pan, a spark ignited the exterior gunpowder and a flame raced through the tube and ignited the interior gunpowder. The powder burned rapidly, creating a large volume of hot gas that blew the ball down the barrel and out.

Usually. Sometimes the spark did not successfully ignite the exterior gunpowder. Sometimes the powder just flashed up without sending the flame down the tube (hence the expression "a flash in the pan").

Back to Mr. Toms, who was suspected of shooting Mr. Culshaw. The pistol had in fact fired, and the ball had raced through the air and lodged in Mr. Culshaw's skull. With it had gone the paper wad that had held the ball in place. The surgeon who removed the ball from the victim's skull also found the paper wad intact and carefully unfolded it. When Toms, who was for other reasons the logical suspect, was arrested and searched, he had a handbill in his pocket. The missing corner of this handbill exactly matched the wad removed from the victim. Mr. Toms went to the gallows.

In 1835 Henry Goddard, then a member of the Bow Street Runners, the precursor organization to the London Police (and later the chief constable of Northamptonshire), noticed that a bullet removed from a murder victim had a small pimple on one side. When he searched a suspect's home he found a bullet mold with a defect in it that would cause bullets cast from it to carry a small pimple on one side. He pointed this out to the suspect, who promptly confessed.

In August 1876 a British police constable named Cock was murdered in Whalley Range, Manchester. Two brothers named Habron were tried for the murder. One was acquitted, but the other, William, was convicted and sentenced to death. His sentence was commuted to "penal servitude for life" a few weeks later, however, as it was felt that he must have had an accomplice (someone other than his brother), and that the accomplice had fired the fatal shot. The day after the Habron trial ended, a man named Arthur Dyson was murdered in Ecclesall, a suburb of Sheffield. The killer was seen and recognized as Charles Peace, an habitual criminal. Peace wasn't caught until two years later when he shot and wounded a police constable named Robinson while attempting a burglary at Blackheath. On his way to the gallows for the Dyson murder, Peace confessed to the murderer of Constable Cock and declared that he had acted alone.

The police didn't want to believe him, but they nevertheless compared the bullets from PC Robinson, Arthur Dyson, and PC Cock. All had been fired from the same—Peace's—gun. The means they used to compare the bullets isn't revealed, but this is one of the earliest known examples of a true ballistics examination. William Habron was released from prison and compensated for the inconvenience of having three years of his life taken from him.

Because of the ubiquitousness of handguns in the United States, and because of the common confusion of propinquity for knowledge, many Americans consider themselves to be firearms experts. Most of them are mistaken. Fifty years ago some of these same misguided people set themselves up as ballistics experts and testified in criminal trials, to the detriment of justice in many cases. In the introduction to his landmark 1935 book, *Firearms Investigation, Identification and Evidence*, J. S. Hatcher observes:

> In cases of death by shooting, it frequently happens that
> the life or liberty of a suspected person may depend entirely

upon the ability of the authorities to determine what kind of weapon did the shooting, or whether a fired cartridge or bullet involved in the fatality did or did not come from a certain gun. The knowledge of firearms and explosives and how they act that is possessed by the public at large, including most peace officers and members of the legal profession, is so meager that almost unbelievable mistakes in testimony vitally affecting the guilt or innocence of an accused person are continually occurring.

During the past twenty or more years, the author has had occasion to observe many instances in which the apprehension of some individual responsible for a crime with firearms or explosives has been delayed or rendered impossible by lack of even a small amount of knowledge of the right kind by the detectives or police who made the investigation. He has observed other instances in which attorneys, courts and juries have been imposed on or misled by insufficiently informed or unscrupulous individuals who have managed to qualify as expert witnesses on firearms.

A spinning top tends to point in the same direction, and, if it is spinning fast enough, it strongly resists any attempt to change its direction. A rapidly spinning gyroscope is so change resistant that it may be used as a compass. Similarly, a football will fly straight and true if the passer imparts a spin to it as he throws. Absent the spin, a football will tumble, yaw, and end up anywhere. Likewise, if you put a point on a bullet and impart a spin to it as it's fired, it has a much better chance of going where it's aimed.

In the eighteenth century, to give the bullet a spin and encourage it to fly straight, gun makers got the idea of cutting spiraling grooves into a gun barrel. These grooves and "lands"—the surfaces left between the grooves—were together called rifling.

The guns that featured rifling became known as rifles. During the American Revolution the Kentucky rifles made by German gunsmiths in Pennsylvania and carried by the frontiersmen were accurate to a far greater distance than the smoothbore muskets used by British troops. George Hanger, a colonel in the British army, became interested in this phenomenon when his bugler's horse was shot out from under him at a distance that Hanger measured several times—a "full 400 yards."

"I have many times asked the American backwoodsman what was the most their best marksmen could do," he reported. "They have constantly told me that an expert marksman, provided he can draw good & true sight, can hit the head of a man at 200 yards." The British army's standard-issue musket, nicknamed "Brown Bess," even under the best conditions was barely accurate at 50 yards.

But a rifle must also have a close fit between bullet and barrel. Otherwise the gases simply blow out through the grooves. And it was difficult to ram a tight-fitting bullet down a muzzle-loading rifle. With the perfection of the breech-loader in the mid-nineteenth century, the musket slowly faded away as breech-loading rifles and rifled handguns came into wide use. The number of rifling grooves in these new guns usually varied from four to seven. The direction of the twist—either right- or left-handed—depended on the whim of the maker.

In 1899 Alexandre Lacassagne, a professor of forensic medicine at the University of Lyons, became the first known person to compare markings on a bullet taken from the body of a murder victim to the rifling found in the handguns belonging to several suspects. He found that the bullet had seven longitudinal grooves created by its passage through a rifled barrel. Only one of the weapons he examined had seven matching grooves. The owner of this gun was duly convicted of murder.

THE STIELOW CASE

CONDEMNED MEN ARE FREED BY A PARDON

Auburn, N.Y., May 9.—Charles F. Stielow, who was pardoned by Governor Whitman after serving three years of a life sentence following his conviction on the charge of murdering Charles B. Phelps and his housekeeper, Margaret Wolcott, at West Shelby, in March 1915, was released from Auburn prison today. He left at once for Buffalo. Nelson Green, who was sentenced to twenty years as an accessory of Stielow and who also was pardoned by the governor, was released at the same time.

—*Altoona Tribune*, May 10, 1918

The scientific investigation of ballistics evidence in the United States could be said to have begun during and because of the 1915 case of Charlie Stielow and Nelson Green. Early on the morning of March 22 Stielow, described as "an illiterate, good-natured tenant farmer," left his house in Orleans County, New York, to discover a dead woman in a nightgown on his doorstep. Footsteps in the fresh snow showed that she had come from the nearby house of Charles B. Phelps, Stielow's boss and the owner of the farm. When Stielow bent over to take a close look at the woman, he recognized Margaret Wolcott, Phelps's housekeeper. She had been shot in the chest (right through the heart, as it turned out). Hurrying over to the Phelps house, he found the kitchen door open and the ninety-year-old Phelps lying on the floor. He had three bullet holes in him and was barely alive. Stielow immediately ran to a neighbor's house and gave the alarm.

The local police, unaccustomed to serious investigation, came to the Phelps house and wandered about, accomplishing nothing constructive and managing to tramp over any forensic evidence that might have been present. The Orleans County authorities hired George W. Newton, a private detective from Buffalo, New

York, to investigate. Ten days later, at Newton's suggestion, the police arrested Nelson Green, Stielow's brother-in-law, who lived with Phelps and Phelps's wife. They took him down to the station house where by two o'clock the next morning Newton had him signing a confession that went into great detail about how he and Stielow had committed the murders. When Stielow was arrested, he duly confessed.

The two men told roughly the same story. Considering that they had been held separately and that they had not actually committed the crime, this says something about the persuasiveness of Detective Newton's interrogation techniques. According to their confessions, they had knocked on the kitchen door. When Phelps came to answer, they shot him and then headed for his bedroom, where they thought he kept his money. Just then Margaret Wolcott came running out of her room and through the kitchen door, slamming the door behind her. Either Green or Stielow (each man claimed the other had done all the shooting) had then fired at the housekeeper through the glass panel in the kitchen door before returning to the bedroom and searching for the money. Each suspect confessed that the two had found about two hundred dollars and that the other man had kept it.

When they returned to their own house, they found a wounded Margaret Wolcott pounding at the front door and screaming to be let in. So they entered at the back of the house and allowed her to scream until she died.

Newton discovered that they had lied when they both denied owning a weapon. Stielow had given both a .22 caliber revolver and a .22 rifle to a relative to hide. The victims had been killed with a .22.

Stielow was tried first and, though he repudiated his confession, saying that it had been coerced, it was admitted into evidence. All that remained was to tie the bullets recovered from the victims' bodies to one of Stielow's guns. The prosecution relied on

Dr. Albert Hamilton, a self-styled firearms expert from Auburn, New York. A patent-medicine hawker who had awarded himself a medical degree, Hamilton had recently become an expert in all things forensic. In addition to firearms, he would render his learned opinion on matters chemical, toxicological, microscopical, and anatomical. He would expertly analyze bloodstains and handwriting as well as conduct autopsies.

"Doctor" Hamilton testified that he had microscopically examined Stielow's revolver and found nine defects at the end of the muzzle that matched nine scratches in each of the four bullets taken from the bodies. When asked why these scratches didn't show up on enlarged photographs of the bullets, he replied that by some mistake, they were of the opposite end of the bullet.

Why, he was asked, would defects at the very end of the barrel mark the bullets? He replied:

> The cylinder fitted so tightly against the rear of the barrel
> that there was no leakage of gas at the breech. The full
> force of the gas following the bullet out at the muzzle, the
> lead expands as it leaves the muzzle, fills in any depressions
> existing at the outer edge of the bore and receives scratches
> from the elevations existing between said depressions.

The jury found Stielow guilty of murder in the first degree, and the judge sentenced him to death in the electric chair. Green pled guilty in order to escape electrocution, and was sentenced to life in prison.

In February 1916 an appeals court upheld the convictions, finding that, "from an examination of the record, it is inconceivable that the jury could have rendered any other verdict."

While awaiting execution at Sing Sing, Stielow somehow managed to convince Deputy Warden Spencer Miller, Jr., that he might be innocent, or at least that his claims warranted attention.

Miller told Louis Seibold, a reporter for the *New York World*, about the case. Seibold in turn hired a Buffalo detective named Thomas O'Grady to investigate. O'Grady discovered that both Stielow and Green were illiterate and that neither was capable of writing his confession or of employing some of the phrases that appeared in the documents. He also discovered that both Newton and Hamilton had worked on a contingency basis: they would not be paid unless Stielow and Green were found guilty.

In the meantime, second and third appeals for a new trial were denied. But the governor had taken an interest in the case, and on December 4, 1916, he commuted Stielow's sentence to life in prison and appointed a Syracuse attorney named George Bond to look into it. Bond drafted Charles Waite, an investigator from the New York attorney general's office, to assist him.

They quickly realized that the confessions of the two men did not correspond with the facts of the case. Both confessions had Margaret Wolcott running right past her two killers on her way out the door. If so, she must have recognized them. Yet she ran up to Stielow's front door to ask for help. And, despite her having been shot through the heart, the men's confessions had her still alive when they emerged from Phelps's bedroom and returned to their house. In their confessions she cries, "Charlie, please let me in, I am dying!" In addition, the alleged angle of the bullet that hit Wolcott as she ran toward Stielow's house was geometrically impossible. Anyone who bothered to visit the scene would have recognized this immediately.

Stielow's .22 revolver was given to experts from the New York Detective Bureau. They examined it before test-firing it in order to judge how long it had been sitting without being fired. "Certainly three or four years, apparently a longer time," was their conclusion. Then they wrapped a sheet of paper around it and fired one bullet. Instead of there being no leakage of gas at the breech as Dr. Hamilton had testified, the paper burst into flames from the heat and power of the discharged gases. Next, bullets were fired into a cotton-filled box and then taken to Dr. Max Poser, an expert in microscopic examination at the Bausch & Lomb headquarters in Rochester. Not only was he unable to find the microscopic scratches that Hamilton had sworn to, but he found that the murder bullets had been fired from a gun with a manufacturing flaw—it lacked one of the five grooves that were supposed to be in the barrel. Stielow's gun could not have fired the bullets that killed Mr. Phelps and his housekeeper. Stielow and Green were promptly pardoned by the governor.

As a result of the Stielow case, Charles Waite realized that the ability to connect a bullet from a crime scene with a specific gun would be of great help in almost every case in which a bullet was collected from a crime scene. And a great step in that direction would be to assemble data on all the guns being manufactured: their bore diameters, the number and width of their lands and grooves, the pitch and direction of their rifling, and anything else that might leave a distinguishing mark on a bullet fired from them. For several years Waite traveled the country collecting data on firearms, poring through old record books, and interviewing the retired foremen of defunct arms companies, many of whom still kept their work notebooks.

At the end of World War I, when a flood of cheaply made foreign pistols came into the country, Waite found it necessary to spend time abroad, mostly in Europe, in order to broaden his information base.

Meanwhile in the 1920s, ballistics investigators Calvin Goddard and Philip Gravelle invented the comparison microscope, a binocular device in which each eyepiece looks through a separate microscope at a different viewing area. An earlier, simpler version was already in use for comparing grains or ground pigments.

An early comparison microscope. The device was invented in the 1920s by Calvin Goddard and Philip Gravelle.

Goddard and Gravelle modified it so that bullets or shells could be compared simultaneously, side by side.

In 1925 Goddard and Gravelle teamed with Waite to establish the Bureau of Forensic Ballistics in New York City, from which they offered their combined expertise to law enforcement agencies around the country. In addition to weapons comparisons, the bureau offered fingerprinting, blood typing, and trace evidence analysis.

On St. Valentine's Day 1929, two men dressed as policemen lined six hoodlums and a passing doctor against the wall of a garage at 2122 North Clark Street in Chicago. Two other men, wearing trench coats and carrying Thompson submachine guns ("Tommy" guns), entered the garage and fired seventy rounds at the lined-up men. Most of them died instantly, and none survived for more than a few hours. The shooters were escorted out by the "policemen" and disappeared. One survivor, a gangster named Frank Gusenberg, when asked who had shot him, managed to gasp, "I'm not gonna talk—nobody shot me." He died on the garage floor with seventeen bullets in him. Hoodlum boss George "Bugs" Moran, the probable target of the killers, arrived too late to be included among the victims. When asked who he thought had done it, he replied, "Only Capone kills like that."

About a year after the massacre, two Thompson submachine guns were found in the home of a known killer named Fred Burke as he was being arrested for the murder of a Michigan policeman. Goddard compared the St. Valentine's Day murder bullets with test bullets fired through these Thompsons and identified the two guns as the murder weapons. Who actually used them in the garage that day is still a matter of speculation.

Over the years firearms examiners have expanded on and refined their techniques so that they can now exploit the many peculiarities of different weapons and their ammunition. The caliber of the bullet, the number and size of the rifling grooves in the barrel, and the position of the marks on the shell are known as "class characteristics." All similar weapons will have the same characteristics. The barrels of all .45 caliber Colt automatic pistols, for instance, will have six rifling grooves with a left-hand twist. The groove depth in the Colt .45 is .0035 inch, and the rate of twist is one full turn in 16 inches.

"Caliber" is the measure of the diameter of the bore (the hole the bullet goes through) in hundredths of an inch—a .30 caliber gun has a bore of 30 one-hundredths of an inch. But the simplicity of the system has become lost over time, and drift occurs. Thus the .38 caliber Colt special has a bore of only .346. And the so-called .38-40 has a bore of .401. And over the years many small-arms manufacturers have made specialty guns of unusual calibers.

The caliber of European guns is measured in millimeters; .30 caliber is, to the Europeans, 7.63 mm. The popular 9 mm caliber is .354 inch and therefore roughly corresponds to our standard .38 caliber, which, because of drift, is actually .359.

If you are examining a bullet under a microscope and you need to determine the rate of twist of the barrel, you must measure the diameter of the bullet and the angle the groove makes relative to a straight line drawn from the point of the bullet to the back. Then you get to use your high school geometry. The formula for determining the twist in inches is $P = \pi \times D \div tan\ a$. P is pitch, D is the diameter of the bullet, and *tan a* is the tangent of the angle of the groove. For example, suppose you're looking at a .45 caliber bullet with a diameter of .451 inch and you find the angle of the groove to be 5° 04'. Your scientific calculator tells you that the tangent of 5° 04' is .0885. You multiply *pi* (3.14159) times the diameter and get 1.4168. Divide that by .0885 and you discover that the twist is one turn in 16 inches.

As you can see from this brief overview, a forensic ballistics expert would need charts, reference books, and a fund of specialized knowledge just to handle the problem of class characteristics.

Individual characteristics complicate matters further. After the expert has examined his charts, stared at the bullet through his microscope, and told you that the round was fired from a Webley-Fosberry .455 automatic revolver, a question remains.

Was it the same Webley-Fosberry that the prime suspect's grandfather brought home after serving with Her Majesty's Own
Twelfth Grenadiers during the Boxer Rebellion? Or was it some
other Webley-Fosberry?

When the bore is reamed out of the solid metal rod that is to
become the gun's barrel, the bore-reaming tool leaves behind a
myriad of tiny scratches. Then comes the smoothing tool, which
reduces these scratches to microscopic size, followed by the tool
that cuts the grooves in the barrel and creates the lands. These
too create their own pattern of scratches. And because each cut
creates minute changes in the cutting tool itself, the metallurgical
structure of each barrel is slightly different. So, even at the microscopic level each barrel leaves a different pattern of striations on
the bullets passing through it. And two bullets fired through the
same barrel in close succession (and before wear and tear, rust,
or mishandling have significantly changed the barrel markings)
should have very similar markings.

CONSTABLE GUTTERIDGE

Shortly after 6 A.M. on September 27, 1927, the still-warm body
of Police Constable George Gutteridge of the Essex County Con-
stabulary was found on the Ongar Road just short of Howe
Green. He had been shot four times, twice under one ear and
once in each eye. These last two bullets were most certainly delivered postmortem. Gutteridge's report book lay at his side, and
he still clutched a pencil stub.

Scotland Yard was called in immediately, and Chief Inspector
James Berrett took charge. Two of the murder bullets had passed
through the constable's body and were retrieved from the road
surface; the other two were gathered at the autopsy. All four were
.45 caliber.

An automobile tire print was found in the grass on the side of the road where Gutteridge had stood. His open report book and the fact that his flashlight was still in his coat pocket led Berrett to conclude that the constable was attempting to write by the light of an automobile headlight when he was shot. Had Gutteridge been writing down the license number of the car? Who would not want their license number written down? When Berrett asked whether any cars had been reported stolen in the area recently, he learned that a Dr. Lovell of the village of Billericay, twelve miles from the scene of the murder, was missing his car.

The murder of a local constable aroused a great deal of anger in the area. Leads poured in, all of which had to be followed up to make certain they were as baseless as Berrett suspected. Several locals suspected the same man, an eccentric who was believed to own a gun. He was quickly cleared. When a man in Basingstoke, sixty miles from the crime scene, confessed, police spent precious time interviewing him only to discover that he had a habit of confessing to murders he had not committed. He was cleared.

Dr. Lovell's missing car was found by police in Brixton, a division in the south of London. Soil and grass found in the wheels on the left side of the car matched the grass on the bank of Ongar Road. Bloodstains identified as human were found on the running board. The car was examined carefully for fingerprints, but no useful ones were found. One spent cartridge case was discovered behind the left front seat.

The bullets retrieved from the crime scene were turned over to experts at the Royal Arsenal, who determined that they were fired from a Colt, a Smith and Wesson, or a Webley revolver. They then passed the bullets on to Robert Churchill to see if he could take the investigation further.

By 1927 Churchill was one of the leading forensic ballistics examiners in Great Britain. A gun maker by trade, he had taken over the business established by his uncle, Edwin Churchill, when

Edwin died in 1910. Robert began his career by giving expert evidence in gun-related cases and in the process formed a friendship with Sir Bernard Spilsbury, the famous Home Office pathologist. This inspired him to expand his knowledge of guns beyond their mechanical characteristics and into bullet and shell comparisons and the other minutiae of forensic ballistics expertise. In 1927, having heard of Goddard and Gravelle's invention, he had a comparison microscope built for his own use.

By looking at the bullets that had killed Constable Gutteridge through the comparison microscope and by further comparing them with rounds fired from various types of revolvers, Churchill declared with some assurance that the gun used in the crime was a Webley service revolver.

Now the police knew what to look for. "Acting on information received," they staked out a garage in Battersea owned by Frederick Guy Browne, a man with a record of burglary, car theft, and violence. He was arrested on January 20, 1928, as he drove into the garage. In the driver's side door pocket the officers found a fully loaded Webley revolver.

The next day Churchill ran comparison tests on bullets fired from Browne's Webley and on the murder bullets. The comparison microscope showed a match, but the murder bullets were badly deformed—the similarity would be difficult for a jury to see. And because no British jury had ever been asked to accept the results of a comparison microscope identification, Churchill wanted his first case to be clear-cut. So he checked the shell casing found in the abandoned car against shells fired from the Webley and found that the breechblock markings on the Webley agreed with the pattern on the shell from the car. Churchill wanted to make it even more positive. He examined the breechblock patterns of the 1,374 handguns placed in the Royal Arsenal for repair and found that not one could be mistaken for the gun that had made the markings on the shell casing. Only then was he ready to go to trial.

On April 23, 1928, Browne and his accomplice, William Henry "Pat" Kennedy, went on trial at the Old Bailey. The forensic evidence clinched their conviction. They were hanged on May 31.

OOPS!

When Virginia police entered the house of Robert and Barbara Parks on February 18, 1950, they found Robert, a thirty-eight-year-old former army captain, in the bedroom. He was dead from a gunshot wound: a bullet had entered his right side, passed through his chest, and lodged behind his heart.

Barbara was twelve years younger than her husband. Robert Parks was known to have a violent temper, and for the past few years their marriage had been rocky. Just a couple of weeks before, Barbara had telephoned a friend in San Francisco in order to borrow the money for a one-way bus ticket. She wanted to leave Robert.

Barbara's story, as she told it through hysterical sobs, was that she was in the kitchen when she heard a shot. Racing into the bedroom, she found Robert standing by the door. "Honey, the gun backfired," he said, then fell dead.

The gun in question, an automatic pistol, lay against the far wall of the dining room. One shot had been fired, and the cartridge case had not ejected properly—it was jammed in the ejection port. By the angle of entry of the bullet into Parks's body and the lack of powder stippling, the forensic investigators concluded that Parks could not have been holding the gun himself when it was fired. Barbara Parks was subsequently arrested for her husband's murder.

But investigators had two questions to answer before winding up the case: Why would Barbara Parks tell such an improbable story when more plausible explanations were available to her? She might have claimed self-defense, or that she had mistaken her husband for a burglar, or that he was teaching her how to shoot when the gun discharged by accident. And how did the hot-air grille in the floor between the bedroom and the dining room get a dent that chipped the paint down to the metal?

The detectives wrapped up the evidence—the gun, bullet, cartridge case, and hot-air grille—and sent them to the FBI Crime Laboratory where technicians verified that the gun had indeed fired that particular bullet, which in fact had been encased in that cartridge. They examined the hot-air grille and saw that the dent in it matched two points on the slide and hammer of the automatic. Under a microscope they observed tiny flecks of paint from the grille at just those points on the gun.

As FBI technicians reconstructed the event, Parks threw the gun down in a fit of temper; the gun hit the hot-air grille at just the wrong spot and then fired. The bullet hit Parks—the angle of the entry wound was just right—and the cartridge case jammed in the ejection port as the pistol's recoil bounced the gun along the floor.

The FBI sent a firearms expert to Virginia to testify to these findings. The judge ruled that Parks's death was an accident. Barbara Parks's incredible description of the events was in fact true. She was released from jail.

The Integrated Ballistics Identification System, or IBIS, is a system developed by Forensic Technology, a Canadian company, for the purpose of analyzing, storing, and sharing bullet comparisons and other ballistics information. In 1993 it was bought by the U.S. Bureau of Alcohol, Tobacco, and Firearms (ATF) at the same

time the FBI was working on a system of its own called Drugfire. Within five years the FBI abandoned Drugfire and worked
out a deal with ATF for the shared responsibility of IBIS. Under
their agreement, the FBI would provide and maintain the necessary communications network, while ATF would manage field
operations.

IBIS is a sophisticated, many-faceted system. Using special
IBIS equipment, the forensic examiner makes photographs of bullets or shell casings retrieved from a crime scene and enters them
into the IBIS database along with background information—caliber, date of crime, date of entry, and rifling specifications. Using
complex mathematical algorithms, IBIS then compares the newly
entered data with its stored information. Candidates for matches
are examined on a computer screen by a forensic examiner. If a
comparison looks like a possible match, the actual physical evidence is then compared under a microscope as a final verifica-
tion. An identification that links two or more different crimes is
considered a "hit" and is assigned a special designation for future
reference.

The National Integrated Ballistic Information Network pro-
gram (NIBIN) has cross-connected IBIS into more than two hundred crime labs across the United States. Further, the IBIS system is used in over thirty countries worldwide. It is credited with
developing thousands of leads that otherwise might have been
missed.

In its quest to wring every possible bit of information from the
bullets found at a crime scene, the FBI also developed a technique
known as CABL, the Compositional Analysis of Bullet Lead. The
theory was simple: most bullets are made of lead, or contain at
least a core of lead inside a harder metal jacket. This lead contains impurities, elements that remain when the lead is refined or
cast, as well as elements deliberately added to change the lead's

properties. (Antimony, for example, may be added to increase the lead's hardness.) Furthermore, lead derived from ores mined in different locations will contain different impurities and in varying ratios. The additives of different bullet manufacturers will also vary in type and amount.

So, if you wish to connect a crime-scene bullet with a suspect and find that the bullet is too damaged or fragmented to compare it with a bullet fired from the suspect's weapon, you can check the composition of the lead and compare it to the lead from bullets found in the suspect's possession. If they are identical or very close, the FBI reasoned, the bullets probably came from the same batch of lead and possibly even from the same box of ammunition.

When the FBI began analyzing bullet lead in the 1970s, only three impurities were measured—antimony, arsenic, and copper. In 1990 more sensitive tests came into use, and tin, bismuth, silver, and cadmium were added to the list of detectable elements. The FBI believed that these tests were valid and useful, and their forensics experts testified to that effect in criminal trials. Like this, for example:

> We can't tell them apart. That tells us that they were
> manufactured or that they were likely manufactured in the
> same pot of lead at a bullet manufacturer. So out of the
> whole population of nine billion or so cartridges that are
> produced here in the United States, we can narrow it down
> to tens of thousands of bullets being produced that would
> have the same composition.

When over the years the FBI itself began to question these assumptions, it asked the National Academies to check their scientific validity. The Academies issued a report, *Forensic Analysis: Weighing Bullet Lead Evidence*, suggesting that the FBI had been perhaps too eager and that its conclusions were more positive than was warranted by the evidence. The Academies discovered, for instance, that bullet manufacturers regularly dump lead left over from one batch of bullet making back into the cauldron used for a new batch, significantly altering its chemical characteristics. Similarly, the castings of new bullets are dropped into vast bins before insertion into shells, so that bullets from one batch are easily mixed with bullets from another. Further, the chemical composition of a mass of molten lead can vary from point to point within that mass. And no consideration was being given to packaging ammunition from the same batches together; the report noted that the FBI identified bullets from fourteen different sources in one individual box. The report concluded:

> CABL does not . . . have the unique specificity of techniques such as DNA typing to be used as stand-alone evidence. It is important that criminal justice professionals and juries understand the capabilities as well as the significant limitations of this forensic technique.

8 : The Footprints of a Gigantic Hound

THE ANALYSIS OF TRACE EVIDENCE is one of the most useful as well as one of the most disputed corners of forensic investigation: useful because it deals with what the criminal may have left behind or taken away with him from the crime scene; disputed because, though couched in the trappings of science, much of it is based on anecdote or "common sense" and has never been rigorously tested. As Albert Einstein put it, "Common sense is the collection of prejudices acquired by age eighteen." Sometimes these anecdotes are valid generalizations and sometimes the common sense turns out to be valid science, but there is no way to know which is which without performing validating tests.

When any forensic procedure becomes the old, established way of doing things, and evidence derived from it is admitted in criminal cases without question, error may creep in. And when the procedure in question is based on unsubstantiated science, the error may be severe and damaging. Not all, or even most, trace evidence is invalid, but the possibility is something to be alert to.

Trace evidence consists of those small, possibly microscopic, bits of detritus that the criminal leaves behind at the crime scene or takes with him. It can be collected and analyzed to establish a connection between the criminal and the crime. Among the possibilities are blood (and other body fluids); explosive residue; animal, vegetable, or artificial fibers; glass shards; hairs; paint scrapings; shoe or foot impressions; tire impressions; bite marks (discussed in Chapter 15); markings on bullets or cartridges; lip prints; ear prints; glove prints; bits of food or other debris; and scrapings of anything from which material may have been transferred.

Dr. Edmond Locard (1877–1966) was the author of the important seven-volume *Traité de Criminalistique* (1923) and the first man to use the term "criminalistics." A French criminalist, Locard held degrees in both medicine and law, and was a student of Alexandre Lacassagne, the professor of forensic medicine at the University of Lyon. Locard became Lacassagne's assistant but left in 1910 to establish his own police laboratory. His enterprise eventually became the official laboratory of the technical police for the Prefecture of the Rhône. His skills were so highly regarded that he was in constant demand by police agencies throughout Europe.

In 1910 Locard stated what is now known as Locard's Exchange Principle: "Any human action, and certainly the violent action of committing a crime, cannot occur without leaving a trace."

In 1912 Locard had a chance to demonstrate his theory when he was called in to aid the Lyons police judiciaire in the investigation of the murder of a young woman named Marie Latelle. The girl had been strangled in the parlor of the house she shared with her parents, and there were indistinct marks on her neck, which the police suspected might be fingerprints. From the condition of the body, the police inferred that Marie had been killed early the night before. She had a boyfriend, Emile Gourbin, a bank clerk, who was the logical suspect. His relationship with the lovely Marie had been volatile at best; she liked to flirt, and he was madly jealous. He was arrested, but he had a strong alibi: at the time of the murder he was some kilometers away dining with friends. These friends not only supported his alibi, they testified that he had stayed downstairs playing cards until about one in the morning and had then retired to an upstairs bedroom to spend the night.

The police judiciaire called Locard to see if he could help—perhaps he knew of some way to develop the fingerprints on the victim's neck. Locard examined the body. He peered closely at the neck through a magnifying glass. He was sorry, he told the police, but the marks on the neck were not fingerprints. They were scrape marks, probably from fingernails.

Locard went to the jail and took scrapings from under the suspect's fingernails. Back in his lab he peered at the material through his microscope. There were skin cells that might have been the victim's, but they might also have been Gourbin's. In addition there were tiny grains of something else, something pink. Taking a close look at this something else, Locard found it to be made up of rice powder containing small amounts of zinc oxide, magnesium stearate, and a form of iron oxide known as Venetian red. Slightly darker than scarlet, Venetian red pigment was used in expensive face powders along with rice powder, zinc oxide, and magnesium stearate.

The police questioned all local compounders of face powders—there were no mass-market cosmetics back then—and located one who produced a product identical to the one found under Gourbin's fingernails. They had, indeed, sold their powder to Marie Latelle.

When confronted with this evidence, Gourbin confessed. To provide himself with an alibi, he had set the clock at his friend's house an hour and a half ahead. Then he had sneaked out to meet Marie. The assignation had been for quite another purpose, he assured them, but when he and Marie had a fight, he strangled her. Gourbin's confession was one of the earliest successes of Locard's use of microscopy and trace evidence.

Locard also contributed to the developing study of dactyloscopy (fingerprint analysis), and invented poroscopy, the study of pore patterns on the papillary ridges of the fingerprints as an aid in identification. In 1914 Locard published the results of a statistical analysis of fingerprint patterns and established rules for comparing prints that still hold true today. In establishing points that match between a suspect print and the exemplar:

—If more than twelve points concur and the fingerprint is sharp, the two prints match.

—If eight to twelve points concur, the case is borderline and the certainty of the match will depend upon the sharpness of the fingerprints; the rarity of its type; the presence of the core of the print and a delta in the comparable part of the print; the visibility of pores; the clarity and obvious agreement of the width of the papillary ridges and valleys, the direction of the lines, and the angles of the bifurcations.

—If fewer than eight points concur, the fingerprints cannot be used for positive identification but only to say that the suspect cannot be excluded.

Locard was one of the founders of the International Academy of Criminalistics in 1929, a model for today's forensic science organizations.

On the morning of Good Friday, April 10, 1936, thirty-four-year-old Nancy Evans Titterton was raped and murdered in her apartment at 22 Beekman Place in Manhattan. She was found by Theodore Kruger, owner of an upholstery shop, and his assistant, John Fiorenza, who were returning a couch to the apartment. The front door to the apartment was ajar, so they entered and called for Mrs. Titterton. They found her body in the bathroom, lying prone in the empty bathtub.

This was a high-profile crime. Nancy Titterton was a noted author and book reviewer, and the wife of the literary critic Lewis Titterton, the head of the literary rights department at NBC. Assistant Chief Inspector John Lyons was assigned to direct the investigation and given a force of sixty-five detectives in the hope that he could clear the case quickly.

Titterton's body was naked except for a pair of rolled-down stockings. Her tied-together pajamas had been used to strangle her. She had been raped. A knife that had been wiped clean of fingerprints lay on the bathroom floor. There were signs of a struggle in the bedroom. Ligature marks on the victim's wrists indicated that she had been tied up before she was raped, but the rope had apparently been cut off and taken away. But then Lyons had a piece of luck: when the medical examiner lifted the body, he uncovered a thirteen-inch-long piece of rope, cleanly cut at both ends. It was similar to a Venetian blind cord, but it had not been removed from any of the blinds in the apartment.

There were several apparent leads. A strange man, described as young, pleasant, and well-spoken, had knocked at a downstairs door twice, the last time at three o'clock in the morning, to inquire after two different women, neither of whom lived in the building. But the man could not be located, and none of the other leads went anywhere. Lyons decided to see if science could help. As the *New York Times* put it: "[The police] made no concealment of their reliance on laboratory examinations of physical items of evidence to lead them to the solution. Fields of science almost ignored heretofore were entered yesterday, and some of the methods of scientific investigation suggested in the Hauptmann case have been resorted to now." (Bruno Richard Hauptmann had recently been tried and convicted for the kidnapping and murder of the Lindbergh baby.)

Two items of possible forensic interest were closely examined. The first was the length of rope discovered under the victim. It was found to have a Tampico fiber base, a product made from the Mexican agave rigida plant and rare enough that it might be traceable. The second was a half-inch-long strand of stiff white hair that was found after careful examination of the bedspread with a magnifying glass. Dr. Alexander Gettler of the New York City Toxicological Laboratory, who found the hair, placed it under a microscope and decided it was horsehair. Furniture was often stuffed with horsehair. Lyons got a sample of the hair from the couch the workmen had delivered, and Gettler found the two to be identical. Lyons then had his men check all the other furniture in the apartment and found that none of it was stuffed with anything that matched the hair found on the bedspread. So the hair had come from the couch. Perhaps it had been carried in on the clothing of one of the furniture repair men.

Theodore Kruger, owner of the upholstery shop, told Lyons that Fiorenza, his assistant, had been with him when they had picked up the couch the day before the murder, and had not come to work until almost noon on the day of the murder. When Lyons pulled Fiorenza's rap sheet he found that in his twenty-four years Fiorenza had been arrested four times for grand theft and had served a two-year jail sentence. There was also a 1934 psychiatric report that diagnosed Fiorenza as delusional and prone to wild fantasies.

The upholstery shop assistant was now certainly a prime suspect, but there was not enough evidence to take to a jury. The needed evidence arrived April 17 when the thirteen-inch Tampico fiber cord found beneath the victim was traced to the Hanover Cordage Company of York, Pennsylvania. Their wholesalers had made many sales in the New York area, and one of them had been to Kruger's upholstery shop.

The white horsehair might have found its way into the bedroom innocently, though the two men had not entered the bedroom when they attempted to deliver the couch. But the rope found under the body tied Fiorenza to the crime.

Fiorenza was brought in for intensive questioning. And when he heard that the rope had been traced to the shop, he confessed. He claimed to have returned to the apartment that morning convinced that Nancy Titterton had fallen for him during their brief encounter the day before. When she rebuffed him, he became so furious that he tied her up and raped her. There was a certain superficial plausibility to this, as Fiorenza had been diagnosed as delusional. But if so, why had he brought the knife and the cord?

Afterward he had strangled her and left her in the bathtub. He claimed that she was still breathing when he left her. But if he intended to leave her alive, why did he strangle her?

His story engendered neither sympathy nor belief from the jury that convicted him. He was executed on January 22, 1937.

"I hate this 'crime doesn't pay' stuff. Crime in the U.S. is perhaps one of the biggest businesses in the world today."
—Paul Leland Kirk

In 1929 Paul Leland Kirk (1902–1970) was appointed professor of chemistry at the University of California, Berkeley. Except for the World War II years, he spent his entire career at Berkeley, teaching and developing the techniques of crime-solving that became known as "criminalistics," a term that Kirk himself popularized. He wanted to establish the field not only as a profession but as a recognized scientific discipline. The clearest explication of his beliefs is a statement he made in 1974:

Wherever he steps, whatever he touches, whatever he leaves, even unconsciously, will serve as a silent witness against him. Not only his fingerprints or his footprints, but his hair, the fibers from his clothes, the glass he breaks, the tool mark he leaves, the paint he scratches, the blood or semen he deposits or collects. All of these and more, bear mute witness against him. This is evidence that does not forget. It is not confused by the excitement of the moment. It is not absent because human witnesses are. It is factual evidence. Physical evidence cannot be wrong, it cannot perjure itself, it cannot be wholly absent. Only human failure to find it, study and understand it, can diminish its value.

Kirk's training as a microchemist brought him to the field of forensic investigation, but he was best known as a blood-spatter expert and for his analysis of the crime scene in the Sam Sheppard case (see Chapter 11).

The use of blood-spatter evidence was brought to the attention of forensic analysts and the general public with the Sheppard case, but the idea had been around for quite a while. In the 1890s Dr. Eduard Piotrowski of the Institute for Forensic Medicine in Poland experimented with using hammers and hatchets on the heads of rabbits. He published his findings in the 1895 pamphlet "Concerning the Origin, Shape, Direction and Distribution of the Bloodstains Following Head Wounds Caused by Blows." Piotrowski believed that "It is of the highest importance to the field of forensic medicine to give the fullest attention to bloodstains found at the scene of a crime because they can throw light on a murder and provide an explanation for the essential moments of the incident."

Biological evidence—blood, saliva, and semen samples—is collected as soon as possible at the crime scene. This is not always as simple as it sounds. Dried blood can be red, brown, black, or even green or yellow. And not all red stains are blood. Semen and especially saliva stains are easy to miss. Careful and close examination of the area with an alternate light source, one that uses a part of the spectrum at which the biological samples are more visible, is essential.

Shoe marks or tire impressions may be either flat (a footprint on a dusty floor) or plastic (a tire track in the mud). They are first photographed, rendering what is called a "latent print." In the case of a flat print, it is then lifted with tape like a fingerprint. A suspect's shoe can be compared with the print using side-by-side examination, or with a transparent overlay. A plastic impression will be preserved by making a plaster cast. The same analysts often handle both foot marks and tire marks, and whatever else that comes along for which there is not a specific expert. The examiner will look first for class characteristics: tires with the same tread pattern, shoes with the same distinctive heel and sole markings. Then he or she will look for the "accidentals"—wear patterns, cuts or abrasions that show up in the impression, or substances adhering to the tire or shoe. The class markings will lead the investigator to the make and model of a tire—there are directories that contain this information—but the accidentals will tie the impression to a specific shoe or tire.

Sometimes class characteristics can be very suggestive. There were footprints in the blood at the scene of the Nicole Brown Simpson–Ronald Goldman double murder in Los Angeles. FBI examiner William Bodziak was able to show that they were made by a pair of size twelve Bruno Magli shoes, made exclusively by an Italian company with very limited production. O. J. Simpson, Nicole's ex-husband, was tried for the murders and acquitted. Although he denied ever owning Bruno Magli shoes, a photograph of him wearing a pair was found. The shoes themselves had disappeared. Finding the shoes and showing that the accidentals were a match would have been a powerful piece of evidence against him. On the other hand, if the accidentals had not matched it would have been easier to believe in his innocence.

In order to make use of footprint or tire track evidence, the examiner needs to be able to show exactly what shoe or tire made the mark. Then detectives need to be able to link a specific person with that mark.

SWGTread, a professional organization for footwear and tire examiners, suggests that their members avoid common terminology such as "consistent with" or "responsible for," and use the following standardized terms:

—"identification" (definite conclusion of identity)

—"probably made" (very high degree of association)

—"could have made" (significant association of multiple class characteristics)

—"inconclusive" (limited association of some characteristics)

—"probably did not make" (very high degree of nonassociation)

—"elimination" (definite exclusion)

—"unsuitable" (lacks sufficient detail for a meaningful comparison)

Louise M. Robbins (1928–1987) was a professor of anthropology at the University of North Carolina at Greensboro and the author of *Footprints: Collection, Analysis and Interpretation*. She could discern what others could not and would gladly write it up in a scientific journal or swear to it in a court of law. In 1978, when anthropologist Mary Leakey and her team discovered 3.6-million-year-old hominid footprints in an ancient stream bed at Laetoli in Tanzania, Robbins affirmed that they were human, and further that they were the prints of a man and a woman walking together, and that the woman was pregnant.

In criminal cases where a shoe print was left at the crime scene but there was no shoe to match it to, the prosecution would have Robbins perform a "wear pattern analysis." As she testified time and again, no two people have identically shaped feet or identical gaits. So if she inspected the inside of the defendant's shoes, she could match foot to footprint, even if the print was made by a different shoe. Defense attorneys and many of her fellow scientists took to calling her method the "Cinderella analysis."

In an article published by the Center for Public Integrity, Steve Weinberg wrote:

> Prosecutors around the nation used to retain the services of a University of North Carolina–Greensboro anthropology professor named Louise Robbins, who said she could match crime-scene footprints to the footwear of perpetrators. Few other forensic scientists endorsed the validity of Robbins' techniques. But prosecutors called on Robbins over and over, banking on the good will of the trial judge to certify her as an expert. Robbins helped convict defendants across the nation until her technique was shown to yield results that were no better than chance would have produced.

Robbins began her Cinderella testifying in 1976, and by 1986 she had been a prosecution witness in ten states and Canada. In 1987 the American Academy of Forensic Sciences had Dr. Robbins's cases and conclusions reviewed by a panel of 135 anthropologists, forensic scientists, and lawyers. They concluded that her work had no scientific merit. One member of the panel, the law professor Melvin Lewis, called it "complete hogwash."

Why would judges, who have the responsibility of keeping junk science out of the courtroom, accept the testimony of this glib charlatan? One reason is that she testified for the prosecution. Most criminal court judges, having come up through the prosecutorial ranks, look less critically than they should at prosecutorial evidence.

Just as a specific gun leaves its imprint on a bullet passing through the barrel, so a specific tool—a chisel, screwdriver, or hammer used to pry or chip—will leave its imprint on the deformed surface. And each tool has unique markings that can be compared to markings left behind.

The forensic analyst who specializes in this area of study is known as a firearms and toolmark examiner. These analysts have their own professional organization, the Association of Firearm and Toolmark Examiners (AFTE). Their code of ethics is very strict and covers the many possibilities that might arise in an investigation or a trial, but essentially the examiner is required to be impartial. He is not to take sides, slant testimony, or give more weight to his testimony than the evidence allows.

The three possible conclusions of a toolmark examination are identification—this tool made this mark; exclusion—this tool did not make this mark; and no conclusion—there isn't enough information to know whether or not this tool made this mark.

As in other fields of forensic analysis, toolmark analysis necessarily suffers from unconscious examiner bias. In a recent case a highly regarded ballistics expert who was trying to match bullet fragments to a particular gun was heard to say on turning in his report, "I really had to work hard to make this match, but I finally made it." He knew what he was supposed to find, and he worked hard at finding it. What would he have found if he had not known what he was expected to find?

It is surprisingly common to find an ear print at the scene of a crime. Someone who is hiding and pressed against a wall, or a person who is listening at a window, may well leave the print of an ear behind. And there are forensic scientists ready to swear that they can tell one ear from another by its print mark. The scientific community does not agree, but ear-print evidence still makes its way into the courtroom, sometimes with ill results.

On Tuesday, May 7, 1996, Dorothy Wood, a ninety-four-year-old retired health worker, was smothered to death in her house in Huddersfield, West Yorkshire, England. Police believed it was the work of a burglar who had climbed in through a transom window above her bed. The killer had apparently listened at the window before entering, because clear ear prints were found on the glass.

The police had one suspect, a twenty-four-year-old local resident named Mark Dallagher, who had prior arrests for burglary. Dallagher claimed he had spent the entire night with his girlfriend. She could not verify this, however, because she had taken a sleeping pill. The detectives took impressions of his ears and sent them, along with the ear prints from the window, to Cornelius Van der Lugt, a Dutch ear-print expert. Van der Lugt examined the ear prints and reported that they did not match.

Then Van der Lugt changed his mind and decided that they did match after all. Perhaps it was because the prosecutor had decided to charge Dallagher with the murder and needed Van der Lugt's testimony. The ear evidence was effectively all they had, and no one need know about the first report. By the time the trial began in the Leeds Crown Court, the prosecution had managed to find a jailhouse snitch who would testify that Dallagher had confessed to him while they shared a cell.

The prosecution put two ear-print experts on the stand—Peter Vanezis, the Regis Professor of Forensic Medicine and Science at the University of Glasgow, who said that the ear print might be Dallagher's but that he couldn't swear to it, and Cornelius Van der Lugt, who was now 100 percent certain that the print on the window had been left by Dallagher's left ear. On December 15, 1998, Mark Dallagher was convicted of murder and sentenced to life in prison.

This was a victory for prosecutors and a first in the United Kingdom. One of them said in a BBC interview, "In planning to use the ear-print evidence we sought the advice of experts in order to prove that it could not have belonged to anyone else."

The verdict was set aside on July 25, 2002, by the Court of Appeals on the grounds that it was "unsafe." Dallagher's new lawyers had called on two experts of their own, Dr. Christopher

Champod, an analyst with the forensic science service in Britain, and Professor P. J. Von Koppen of the University of Antwerp in Belgium. Both had made a study of ear prints and both, for slightly different reasons, had found that they were not a dependable means of identification. There would have to be another trial.

Ten days into Dallagher's new trial Van der Lugt's original report surfaced. In it he had stated that the ear print had definitely *not* come from Dallagher. The judge stopped the trial and ordered Dallagher released from prison on bail while he pondered what to do. A short while later a DNA analysis of the ear print itself showed that the donor could not have been Dallagher. Judge Sir Stephen Mitchell directed a verdict of not guilty and said to Dallagher, "This most unfortunate saga at long last comes to an end."

During the same period several convictions in the United States that had depended on latent ear-print analysis, including two at which Van der Lugt had served as the expert witness, were reversed on findings that the technique was unreliable.

Hairs are one of the most common forms of trace evidence found at crime scenes. Like dogs, cats, rabbits, and other animals that might wander across a crime scene, humans are continually shedding hairs. The hair of different species can be reliably differentiated under a microscope, and the hair of two different people might be different enough for exclusion purposes. The practice of bleaching, dyeing, curling, flattening, and otherwise altering hair can further individualize a sample, as can a variety of diseases and parasites. But the error rate for the microscopic analysis of hair is too high to use it for reliably identifying anyone as the donor of a given sample. After DNA analysis became available, the FBI did a study to check their results and found that microscopic examination of hair by their own experts was wrong 11 percent of the time.

DNA can be extracted from the root of a hair, but hair that falls out naturally is usually rootless. So microscopic examination is often still necessary.

On April 19, 1989, a twenty-eight-year-old woman who was jogging in Manhattan's Central Park was attacked, raped, savagely beaten, and left for dead. When she was found five hours later she had lost 75 percent of her blood, her skull was fractured, and one eye was out of its socket. The immediate suspects were a gang of fourteen- and fifteen-year-old boys who had been running wild through the park, knocking people off bicycles, assaulting people, and otherwise causing havoc for several hours around the time of the attack. They had already been rounded up by the police and were being held in the Central Park precinct house when the jogger was found.

The woman survived, though she was unconscious for twelve days and in and out of delirium for another five weeks. She remembered nothing of the attacks, and even years later the entire five-week period was erased from her memory.

Three hairs found on the clothing of one of the boys and one hair found on another were examined microscopically and found to be consistent with the victim's hair. On the basis of this and videotaped confessions that were recanted almost immediately, five of the boys were placed on trial and found guilty.

Three months after the jogger was attacked, eighteen-year-old Matias Reyes was arrested for rape. As part of a plea bargain, he confessed to another rape as well as the rape-murder of a pregnant mother. Thirteen years later, in January 2002, after the statute of limitations had run out, he added the attack on the Central Park Jogger to his list of crimes. Recent advances in DNA testing made it possible to check his story, and semen recovered from the jogger's body was tested. It matched Reyes's. Further, DNA tests on the hairs recovered from the five boys did not match the victim's. At the request of Manhattan District Attorney Robert Morgenthau, the convictions against the five boys were dismissed.

9 : He's Dead Jim

> An autopsy is the systematic and internal examination of a body to establish the presence or absence of disease by gross and microscopic examination of body tissues. . . . Medicolegal autopsies are conducted to determine the cause of death; assist with the determination of the manner of death as natural, suicide, homicide, or accident; collect medical evidence that may be useful for public health or the courts; and develop information that may be useful for reconstructing how the person received a fatal injury.
> —*Strengthening Forensic Science in the United States: A Path Forward*, 2009

THE FIELD of forensic medicine covers everything of a medical nature that might be discussed or argued in a court of law. The responsibilities of a medical examiner are myriad, but generally speaking he or she is charged with determining the cause of death, the time of death, and the identity of the decedent.

The specialized knowledge required to answer these often complex questions is immense. Forensic pathologists are responsible for determining the cause of death when someone dies suddenly, unexpectedly, or violently; at autopsy they look for the presence or absence of disease, injury, or poisoning; they collect medical evidence such as trace evidence and secretions; they document sexual assault and determine how injuries to the body were inflicted. Some forensic pathologists also have a working knowledge of toxicology, wound ballistics, trace evidence, forensic serology, and DNA technology.

The earliest known person who might be said to have practiced what we think of today as forensic medicine was Imhotep, counselor to the pharaoh Zoser who ruled Egypt five thousand years ago. Imhotep, who was both the chief justice of the kingdom and physician to the pharaoh, was a polymath—an early Egyptian combination of Leonardo da Vinci, Thomas Jefferson, and Louis Pasteur. He earned an impressive collection of titles: Chancellor of the King of Lower Egypt, Doctor, First in Line after the King of Upper Egypt, Administrator of the Great Palace, Hereditary Nobleman, High Priest of Heliopolis, Builder, Chief Carpenter, Chief Sculptor, and Maker of Vases in Chief.

Imhotep designed the first Egyptian pyramid, the so-called step pyramid at Saqqara. He improved the manufacture of papyrus scrolls and wrote the earliest known treatise on the surgical treatment of traumatic injuries. The Edwin Smith Papyrus, written a thousand years after Imhotep but believed to be based on his work, lists forty-eight different traumatic injuries and gives advice on their treatment and prognosis. It is notable for its pragmatic, nonmagical approach—only a few magical incantations are called for. In it, Imhotep wrote of the heartbeat in the extremities of the body and of taking a pulse, but mainly he concerned himself with injuries—wounds that might be acquired in combat. A few sample case headings are: a wound in the head penetrating to the bone; a gaping wound in the head with a compound comminuted fracture of the skull; a gaping wound at the top of the eyebrow, perforating to the bone.

In each case, instructions are given as to how to diagnose the wound and then how best to treat it. Some sad cases are labeled "An ailment not to be treated."

The Romans gave some thought to the problems of forensic medicine, and due to the necessities of the battlefield their physicians were skilled in the treatment of wounds. When Julius Caesar was assassinated in 44 B.C., the physician Antistius determined that of the twenty-three stab wounds Caesar had suffered, only the second one had been decidedly fatal. The Romans also had some knowledge of poisons due to the popularity of their use among the nobility.

The Code of Justinian, first issued in 529 A.D. and revised in 534, specified that a medical expert at a trial should not appear for either side but should be appointed to assist the judge impartially—a good idea that might be followed with profit today. Later, in the sixth century, St. Gregory of Tours wrote *A History of the Franks* in which he confirmed that physicians were indeed often called as expert witnesses by the courts.

In 1209 Pope Innocent III accepted testimony from doctors in an ecclesiastical court as to the lethality of a particular wound. This established an important precedent that eventually opened all European courts to such testimony. In 1497 Dr. Hieronymus Brunschwig of Strasbourg published the first known description of gunshot wounds.

In 1532 the Holy Roman Emperor Charles V devised the Caroline Code of criminal law, which advised that a medical doctor should be consulted in all deaths of a violent or unnatural nature. These included death by wounding, poisoning, hanging, drowning, murder, manslaughter, infanticide, abortion, and many other circumstances involving injury to the person. Doctors were required to testify in cases of malpractice and to keep written records of their autopsy findings.

In 1547 Antonius Blancus wrote *On the Indications of Homicide*, in which he questioned the reliability of a Germanic tribal custom that was still in use as a method to establish a murderer's guilt. Known as *ius cruentationis*, it required the suspect to touch the corpse of the victim. If the corpse began to bleed at the touch, the suspect was guilty. Despite Blancus's doubts, the use of the custom continued in some German courts until the mid-eighteenth century.

In 1670 the medical faculty of the University of Prague decided that their expert opinion on the subject of any wound would be given only after the joint consultation of the dean, one professor, three barber-surgeons, and two barbers.

Dr. Theodoric Romeyn Beck, a lecturer on medical jurisprudence in the College of Physicians and Surgeons of Western New York, published the two-volume *Elements of Medical Jurisprudence* in 1823. It covered questions of rape, impotence, sterility, pregnancy and delivery, infanticide and abortion, legitimacy, presumption of survivorship, identity, mental alienation, wounds, poisons, persons found dead, and feigned and disqualifying diseases. For the next half-century the book was the standard reference on forensic medicine. It was republished several times and translated into German and Swedish.

For the power of his deductive abilities, Dr. Joseph Bell (1837–1911) became the model for Arthur Conan Doyle's Sherlock Holmes. A surgeon at the Royal Infirmary in Edinburgh and a lecturer at the University of Edinburgh Medical School (where Conan Doyle received his medical degree), Bell amazed his students with the accuracy of his deductions about patients. And not merely concerning their medical symptoms. "This man is a left-handed cobbler," he once said, startling not only his students but the cobbler in question. "The worn places on his trousers," he continued, "could only have been made by resting a lapstone between his knees. The right side is more worn than the left because he hammers the leather with his left hand."

A story that Bell liked to tell on himself concerned a particular patient that he had decided would provide a good example of his methods. "Here is an interesting case for us," he began. "This man, I should say, is a recently discharged soldier who was probably in the Royal Scots and had a good deal of service in the East. In the Army he was probably in the band, and, I have no doubt played a brass wind instrument."

When his students looked suitably impressed, Bell explained his methods. When the patient entered the room he had stood rigidly to attention. No civilian does that. Neither would a soldier who has been discharged for any length of time. Hence Bell's deduction about the recent discharge. The deep tanning on the patient's face and neck suggested that he served in a hot climate, and the tattoo on his arm suggested it was in the East. His belt buckle was from a Royal Scots regiment. As he was smaller than the usual infantrymen, there was a good chance that he had served in the band. His narrow chest and shallow breathing were the marks of emphysema, possibly caused by playing a large wind instrument for many years.

Bell turned to the patient to verify his deductions.

"Now, my man, have you been a soldier?"

"Yes, sir."

"For a long period?"

"Twenty years, sir."

"You have seen service in India?"

"Yes, sir."

"You played in the band?"

"Yes, sir."

"And can we take it that you played the euphonium or a similar instrument?"

"No, sir. The big drum."

Well, four out of five isn't bad.

The office of *Coronae Custodium Regis*, "Keeper of the King's Pleas," was established in England when King Richard the Lion Hearted was captured by Leopold of Austria in 1192. The original function of the office was to collect taxes, particularly death duties, to be used for Richard's ransom. The functions of the coroner gradually expanded and the tax-collecting aspect disappeared, but the connection with death remained.

"Crowners," as they became known, were charged with keeping track of convicted felons and, if the sentence was death (and almost every felony was a capital offense), with seeing to it that the executed felon's worldly goods were properly confiscated by the crown.

By the time the post of coroner was brought to the United States, its function had narrowed to the investigation of suspicious deaths. There were no special qualifications for holding the job, which was often an elective office. Often the local mortician would be appointed or elected since he was the only one with the facilities to handle dead bodies. The coroner, and in some places a coroner's jury, would hold a legal proceeding known as an inquest or postmortem to determine if a crime had been committed and, if so, whether there was probable cause to say who had committed it. The rules of conduct in the coroner's court were set by the coroner. Witnesses had no right to have an attorney present and might be forced to answer questions that would not be allowed in a formal court of law. The decisions were binding on the victim but not on the suspect, if any. That is, the coroner's findings could be used to settle estate or insurance questions, but it would take an indictment by a grand jury or a preliminary hearing before a trial judge to bring someone to trial on a felony charge.

In New York City in the late nineteenth century, the office of coroner reached a level of corruption and malpractice that was truly creative. Since the coroner was paid a set fee per inquest, he would often hold three or four inquests over the same body. Some coroners let it be known that for an extra $10 they would change the official cause of death from suicide to accidental death, thus assuaging the family's feelings and making it possible for them to collect on the deceased's life insurance. For $50, some would change a finding of homicide to one of accidental death, a cheap enough price to pay to get away with murder.

In 1897 a Brooklyn coroner had the body of a drowned man dragged from spot to spot along the East River waterfront. At each spot he would hold an inquest over the body. Even though the morgue held only one body, he billed the city for $10,000 in fees. The next year, when Brooklyn joined Manhattan and the other three boroughs to officially form the City of New York, the fee system for coroners was done away with and the job was made a regular salaried position. Twenty years later, in 1918, the position of coroner was abolished and the modern medical examiner system was installed.

Shortly before the New York City coroner's office closed, it handled one of its more unusual cases. A husky, six-foot businessman named Murray Hall had died in bed. The coroner's physician determined that the cause of death had been a heart attack and that Hall's gender was actually female, a fact that even Hall's niece, who lived in the same apartment, had not known. The jury's verdict: "We find that Murray Hall came to his death by natural causes. We find that he was a lady."

It takes about two hours to do an uncomplicated autopsy on a person who died of a stroke or a heart attack. Bullet wounds take longer. Mafia killings always take more time

because of the number of bullet holes. We have to check
each injury to see whether it contributed to the death.
                              —Michael M. Baden, *Confessions of a*
                                      Medical Examiner, 1990

The first step in the forensic examination of a body is a positive identification by a family member or close friend. Morgues
handle this in various ways, from the ritual drawing back of the
sheet to New York City's practice of showing a photograph of
the deceased.

Then comes the autopsy, where it is the job of the medical
examiner or one of his assistants to determine the actual cause
of death. Ideally, one of the homicide detectives working the case
is present at the autopsy. This gives the detective a chance to ask
questions and to have them answered immediately and in plain
language. There is a chance too that this will spare the medical
examiner a day in court, since the detective can then testify to
what was found at autopsy.

If the case is a prominent one, the district attorney will probably send an investigator to the autopsy. This is a signal to the
medical examiner that every cut, every assay, and every statement
will be reviewed at trial. Pictures will (or should) be taken at every step of the autopsy. If there are questions afterward (or even
years later), the possibility then exists of obtaining an answer.

On July 4, 1850, Zachary Taylor, four months into his second year as the twenty-second President of the United States, attended the groundbreaking ceremony for the Washington Monu-
ment and fell ill shortly afterward. It was explained that he had
eaten too many cucumbers and cherries while standing in the hot
sun. Five days later, after suffering extreme diarrhea, fever, and
stomach cramps, the sixty-five-year-old president suddenly died.

For political reasons having to do with the slavery question,
Taylor had made many strong enemies in his year and a half in

office. The official cause of his death was gastroenteritis, but for more than 150 years rumors have circulated that he was poisoned, probably with arsenic.

On June 17, 1991, at the request of his closest living relative, Taylor's remains were exhumed and taken to the offices of the chief medical examiner of Kentucky. Samples of his hair, fingernails, and tissue were taken, and he was reburied with appropriate honors.

The samples were subjected to neutron-activation analysis, which revealed the presence of arsenic, but at levels that were hundreds of times lower than they would be if he had been poisoned. Arsenic had probably leached into the body from the soil he was buried in.

But conspiracy theories never die. In his 1999 book *History as Mystery*, Michael Parenti suggests that the neutron-activation analysis was performed incorrectly. He asserts that Taylor was poisoned by arsenic after all.

When President John F. Kennedy was assassinated in 1963, everything about the killing and its aftermath assumed great importance. In the national effort to come to terms with the facts of this terrible event, one of the most critical pieces of information would be the results of the president's autopsy. It was conducted at Bethesda Naval Hospital by three examining pathologists. One of them, Dr. James Humes, a navy commander, wrote in the autopsy report, "The complexity of these fractures and the fragments thus produced tax satisfactory verbal description and are better appreciated in photographs and roentgenograms [x-ray pictures] which are prepared." Unfortunately the FBI agent in charge on the scene decided that the corpsman who was trained to take the pictures had no "clearance" to be present at the autopsy. The only other photographer, an FBI agent who presumably did have clearance, was not trained in photographing postmortem gunshot wounds. As Dr. Michael M. Baden puts it in *Unnatural Death*:

> His pictures showed it. A proper photograph would have shown the injury first as it was and then cleaned off, next to a ruler to give perspective on its size and position in the body. None of his pictures clearly defined the entrance or exit wounds. The photographs of the body's interior were out of focus. You have to know at what level you want to shoot—the chest is deep. He didn't take pictures of any internal organs. These are the pictures Humes proposed to rely on, his own descriptive powers having failed him.

The lack of adequate autopsy photographs is at least part of the reason why the death of President Kennedy remains for many a mystery, and why various conspiracy theories refuse to fade away.

The standard incision for opening the body for internal autopsy examination is in the shape of a Y. It extends from each shoulder to the pit of the stomach and then down and through the pelvis. The internal organs—heart, lungs, spleen, liver, and so on—are removed, weighed, and inspected for physical signs of damage. Tissue samples are then taken and set aside for chemical and toxicological testing. The stomach is removed, its contents examined, and samples taken. The state of digestion of any food present may help determine the time of death.

Any fluid in the thoracic (chest) cavity or in other body cavities is siphoned off and saved for analysis. Any urine present in the bladder will also be preserved for analysis. Drugs that may have been ingested by the decedent can be detected in the urine.

The head is usually examined last. The exterior is closely examined for tiny scratches or wounds, the skull is examined for fractures, and the area around the eyes is examined for petechiae—pinpoint hemorrhages that might indicate strangulation or hanging. The top of the scalp is incised, and a flap is pulled down in front of the face. The skull is then sawed open and the brain is removed for examination. (This is a gruesome and disturbing sight if you are not used to it. But when the section of the skull is replaced and the skin flap is pulled back into place, the face and head can be prepared for a funeral such that the damage is invisible to the viewer.)

The complex and thorough procedures of the medical examiner may seem a waste of time if the cause of death is obvious—say, a gunshot wound or drowning. But many times it has been shown that the seemingly obvious cause of death is only a contributing factor and sometimes not the true cause at all. Even when no surface wound was visible, people have been found to have been shot or stabbed.

If the body was found in the water but there is no water present in the lungs or stomach, the victim was probably dead before entering the water, though there is a condition called laryngospasm, or "dry drowning." In these cases the victim's larynx closes and he is unable to draw water into his lungs. About 10 percent of drownings are dry drownings. If the body was found in the ocean but the water in the lungs is fresh water, the death would most certainly be regarded as suspicious.

If the victim supposedly died in a fire but there are no carbon particles in the lungs and no sign of carbon monoxide poisoning in the system, the fire may have been started to cover up a murder.

When the physical part of the autopsy is completed, the internal organs are placed back inside the body cavity. The body is then sewn back together, the head is reassembled, the face is made to look once more like a face, and the body is placed in cold storage. Chemical and biological tests may take weeks to complete and may be performed at laboratories thousands of miles from the autopsy site. When the cause of death has been officially determined, a death certificate is issued and the body is released to the family.

One of the great myths of forensic science, perpetuated in weekly television shows, is that it is possible to tell the time of death down to the quarter-hour.

"I can say that Mr. Fleishwhacker died sometime between 7 and 8:30 in the morning on Tuesday."

"Can't you pin it down any better than that, Dr. Kildare?"

"Well, you'll be fairly safe if we call it between 7:30 and 7:45."

It certainly would be useful to the homicide detectives to have the time window narrowed down so precisely. And in real life, forensic pathologists do more than their best to oblige. But in truth it is seldom possible to pin down the time of death so closely. And the longer ago the death, the more difficult the task.

The three major markers of time of death are rigor mortis, livor mortis, and algor mortis—the stiffening of the body, the settling of the blood, and the decrease in body temperature. These processes occur at the same time but at different rates. The figures for these rates are given in pathology textbooks and are presented as averages.

At or very shortly after death, all the muscles in the body relax completely. This is called "primary flaccidity." Rigor mortis—the stiffening of the body—usually begins from two to six hours after death. It starts at the eyelids, neck, and jaw and proceeds down the body to the larger muscle groups until all the body's muscles, even the heart, are stiffened. The body stays in rigor for twenty-four to seventy-two hours; then rigor passes in roughly the same order that it arrived.

Livor mortis, also known as postmortem lividity and hypostasis, describes the settling of the blood after death. Since it is no longer being pumped through the body, the blood settles at the lowest points it can reach given the position of the body. This pooling is evident in the skin, where it causes a purplish-red discoloration. It can begin at any time from half an hour to three hours after death. As the blood congeals, it becomes fixed and so can serve as a rough guide to time of death and as an indicator of whether or not the body has been moved. Lividity will always present at the lowest part of the body, so if it is evident throughout the body, it is a sign that the body has been shifted. Often this is due merely to the fact that the paramedics (or whoever found the body) rolled the deceased over in order to check for signs of life.

Algor mortis, Latin for the coolness of death, is the body's path from the average 98.6 degrees Fahrenheit of life to the temperature of inanimate objects. There are different formulas for the rate of cooling, but the commonly accepted one is that the body cools only slightly during the first hour after death, presumably because the metabolic processes are still shutting down. Thereafter it cools at the rate of about one and a half degrees per hour. The temperature of the corpse is best taken rectally or from one of the internal organs. Usually at a crime scene a meat thermometer is inserted into the body on the right side just under the rib cage so that the tip penetrates the liver. It is left in for about five minutes before the reading is taken.

On a hot day in midsummer the temperature of the body might stay the same for quite a while or might even rise. On a very cold day the body temperature might fall comparatively quickly. If the deceased died while in the throes of a high fever, the rate would have to be adjusted accordingly. And there are many factors that can intervene—the deceased's body weight, the amount of physical exertion right before death, and fluctuations in the external temperature—so that the readings are not always dependable. By roughly six hours after death, any temperature readings are absolutely untrustworthy.

Another consideration in determining time of death is that a person who is fatally injured and who should, according to everything we know of medical science, fall dead right then and there might live for quite a while. Thus the time of death and the cause of death are widely separated. In the *Police Journal* in 1943, the British pathologist Sir Sidney Smith told of the case of a man, whom he identified as "an elderly professional," who left his residence hotel in Edinburgh one winter evening and stayed away all night. The next morning at half past seven he returned and was let in by the maid. He was, she noticed, wearing his overcoat and hat and had his umbrella over his arm. His face appeared to be covered with blood. Before she could call anyone he said, "Don't worry. I will just go upstairs and have a wash." After placing his umbrella in the hall stand, and hanging up his hat and coat, he went upstairs to the bathroom. The maid found him there a few minutes later, collapsed and unconscious.

When the police were called, the old gentleman was taken to the hospital, where he died a short while later. The cause of death was a self-inflicted bullet wound to the head. He had held the gun under his chin and fired. The bullet passed through his brain and came out on the left side of the frontal bone of his skull, leaving an exit hole an inch and a quarter in diameter.

By following the blood trail, the police were able to piece together the night's events. At some time before six in the morning, the man had been sitting on a sheltered bench in a garden across the way from his hotel when he shot himself. He then got up and walked in a wide circle for about 165 yards and returned to his bench. Then he rose again and wandered around the garden for a while before once again settling on the bench. Finally he returned to the hotel. A fresh coating of snow on the ground showed clearly that there had been no one else near him, and letters received by relatives the next day put it beyond question that his death was a suicide.

Bits of the gentleman's brain were found clinging to the top of the shelter over the bench. Not knowing the facts of the case, and just seeing the body, any pathologist would be willing to state that the wound would have caused instant unconsciousness and rapid death. But the man had lived, walked, and talked for at least two hours after shooting himself.

10 : Inheritance Powder

I do remember an apothecary—
And hereabouts he dwells—
which late I noted
In tatter'd weeds, with overwhelming
brows, culling of simples. . . .
Noting this penury, to myself I said,
'And if a man did need a poison now,
Whose sale is present death in Mantua,
Here lives a caitiff wretch would sell it him.'
—William Shakespeare, *Romeo and Juliet*

IN 399 B.C. the Athenian gadfly Socrates was sentenced to death for subverting the youth of Athens by encouraging them to question their elders. He carried out the sentence himself by drinking a cup of poison hemlock. As described by witnesses to his death, he expired quite peacefully on his couch.

The next poisoning of historical significance was also self-inflicted. In 31 B.C. Cleopatra, the queen of Egypt, held an asp to her bosom.

The earliest-known professional poisoner was the Roman matron Locusta, a former slave who became an instrument of imperial policy. When Agrippina, wife of the emperor Claudius, decided to do away with her husband, she sought advice about poisons. As the historian Tacitus tells us in *The Annals of Imperial Rome*, Agrippina needed something subtle that would upset the emperor's faculties but defer his death. An expert in such matters was located—Locusta. Her special skills would reprieve her from a recent sentence of death and fit her for a long career of imperial service. After two attempts the desired result was achieved, and in 54 A.D. Agrippina's son Nero became the sixth emperor of Rome. He rewarded his mother by ordering her murder.

In mid-seventeenth-century Europe, poisoning became a popular hobby among the gentry. This was particularly the case among women, for whom traditional methods of slaughter—those requiring physical strength and mastery of edged weapons—were infeasible. The advantages of poison were obvious: it required neither strength nor dexterity, and could be administered in comparative anonymity.

In the 1650s there was a considerable increase in the number of young, rich widows in the larger cities of Europe. Many of them confessed to their priests that they had poisoned their husbands. Bound by the seal of the confessional, the priests could take no action against the offenders. But as they compared notes, the sheer number of these confessions frightened them. In 1659 a delegation of priests informed Pope Alexander VII of the problem.

The Pope, who had temporal as well as spiritual authority over the city of Rome, began an independent investigation. His agents found that a group of young wives, some from among Rome's first families, met regularly at the house of Hieronyma Spara, a reputed witch and fortune-teller. Then, as described by Charles Mackay in *Extraordinary Popular Delusions and the Madness of Crowds*:

> A lady was employed by the government to seek an
> interview with them. She dressed herself out in the most

magnificent style; and having been amply provided with
money, she found but little difficulty, when she had stated
her object, of procuring an audience of La Spara and her
sisterhood. She pretended to be in extreme distress of
mind on account of the infidelities and ill-treatment of
her husband, and implored La Spara to furnish her with
a few drops of the wonderful elixir, the efficacy of which
in sending cruel husbands to "their last long sleep" was
so much vaunted by the ladies of Rome. La Spara fell into
the snare, and sold her some of her "drops" at a price
commensurate with the supposed wealth of the purchaser.

The liquor thus obtained was subjected to an analysis and found
to be, as was suspected, a slow poison—clear, tasteless, and
limpid.

As a result of these investigations, Spara, her assistant Gra-
tiosa, and three young wives were hanged. More than thirty
women were whipped in the streets, and several women whose
social or political positions made it inexpedient to punish them
openly were banished from the city. Over the next few months,
nine more women were hanged and another group flogged.

The practice of poisoning was not restricted to Rome. In 1679
two poison wholesalers, Mesdames Lavoisin and Lavigoreux,
were arrested in Paris. Between thirty and fifty of their retailers,
mostly women, were also apprehended in Paris and in other principal cities of France. The pair were burned alive in the Place de
Greve on February 22, 1680, and their retail agents hanged. This
malevolent band had been the teachers, instigators, and suppliers
for hundreds of poisonings performed overwhelmingly by or on
behalf of women who wished to be rid of their husbands (though
there was a sprinkling of men who sought to eliminate rich fa-
thers or uncles). The poison used, probably a salt of arsenic, was
called by those in the know "inheritance powder."

This wave of seventeenth-century poisonings was attributed to a cabal of women, invariably described as witches or fortune-tellers, who possessed the necessary skills to manufacture the tasteless, undetectable poisons. These women first became the confidantes of their prospective clients, then whispered suggestions in their not-altogether-unwilling ears. That so many women were apparently eager to dispose of their husbands may speak to the impossibility at the time of escaping an unhappy marriage by any means except murder. The husband in a bad marriage could effectively ignore his wife and do just about as he chose. The wife, however, was deprived of her rights by law and custom, and was subject to her husband's command as long as he lived.

Unless the poison was violently corrosive, caused clear physical symptoms that could be observed in the dying person, or remained noticeable on the corpse, there were no techniques at the time for detecting the presence of poison in a body. Those who were caught had probably purchased their potent powders from a supplier who had brought attention on herself. Since torture was still a popular tool of criminal investigation, they were soon identified.

In order to protect society from poisoners, a means had to be found to detect poisons in the body for some time after death. Although the "subtle poison undetectable by science" has long been a staple of detective stories, such a substance is virtually nonexistent today. But the development of reliable means of detection has been a long, slow process.

Dr. Hermann Boerhaave (1668–1738) was the first to describe a practical, if possibly dangerous, way to identify suspected poisons. He advised placing the material in question on red-hot coals. The smoke rising from it was then to be cautiously smelled. A trained nose could probably detect most simple chemical poisons this way, but how safe or reliable this method was is not known. At any rate, it did not catch on.

In *Elementa medicinae et chirurgiae forensis*, Joseph Jacob Plenck (1739–1807) pointed out that the only positive proof of death by poisoning is the chemical identification of the poison itself within the organs of the body. It would be another half-century before this theory was generally accepted.

Modern toxicology, the study of poisonous substances, originated in 1814 with the publication of the first volume of *A Treatise on Poisons Derived from the Mineral, Vegetable and Animal Kingdoms, or General Toxicology Considered with Respect to Physiology, Pathology and Legal Medicine* by Dr. Matthieu Joseph Bonaventura Orfila. It was followed by a second volume a year later. Originally published in French, the book was translated into English, Spanish, and German within only a few years and remained a standard toxicology text for more than a century. Still a student in Paris when he began to catalog poisons, by 1819, when he was only thirty-two years old, Orfila was a professor of legal medicine and within a few years became head of the chemistry department of the University of Paris. In 1840 Dr. Orfela presented evidence at the trial of Marie Lafarge for the poisoning of her husband, one of the most celebrated crimes of the nineteenth century.

Orfela was assisted in his forensic labors by Marie Guillaume Alphonse Devergie, who went on to develop the use of microscopic analysis in forensic pathology. In 1835 Devergie published his findings in his landmark book *Medecine légale, theorique et pratique*.

The classification of objects is often a problem, representing as it does a human need for order rather than an intrinsic relationship of the objects under consideration. So it is with poisons. Authorities may agree that there are three types of poisons, but may then disagree as to what these three types are. One tome may divide poisons by their source—mineral, vegetable, or animal. Another will classify them according to their effects—corrosive, irritating, or systemic. Still another will classify poisons as those which act by attacking the gastrointestinal canal, those which attack the central nervous system, and those which depress the nerves and circulatory systems.

The definition of poison is itself subjective. As the *Merck Manual* says, almost any substance may be toxic if inhaled, ingested, or otherwise taken into the body in the right proportions and under the right circumstances. Water, salt, and even air can be deadly. Some substances, such as the nerve poison curare, are toxic in small quantities and yet are used medicinally in even smaller quantities. Medicines used to cure or ameliorate one condition can kill if prescribed for a different condition. Botulism toxin, the second-deadliest poison known, is now injected into the facial muscles of members of our aging population to eliminate wrinkles. Insulin, taken daily by diabetics to replace the natural insulin no longer manufactured by their bodies, will kill if administered in a large dose. Several chemicals used as anesthetics during surgery will kill subtly, quickly, and painlessly. These break down in the body so as to be essentially undetectable at autopsy unless their use is suspected. Even then they are very difficult to detect. So whether or not a substance is a poison sometimes depends on the intent of the user.

The medical examiner today has a variety of sensitive machines available to help determine if poison is present in a sample of tissue or fluid. But every body cannot be routinely examined for the presence of poison. Many poisons leave no outward sign, and the signs left by others mimic the appearance of natural causes of death. Still, poisoners often give themselves away through overconfidence, greed, and a tendency to murder again at the slightest provocation. It is true that poisoners often suffer from a fatal hubris. Unfortunately the poisoner's hubris is usually fatal to several innocent people before he himself is caught.

The subject of poisons and poisoning has been rife with misconceptions and myths. For instance, it was once believed that boiling a food suspected of containing poison and then dropping a silver coin into it would remove the poison. Alternatively, it was thought that the coin would turn black, thus indicating the presence of poison. This is incorrect. Silver turns black in the presence of sulfur, which is not usually used as a poison. There was also a common belief that the corpse of a person who had been poisoned would turn black and blue, appear spotted, or smell bad—a description of the normal processes of decomposition.

Until about two hundred years ago it was widely believed that tomatoes were poisonous, possibly because they are a member of the deadly nightshade family of plants. There is an unverified story that a British agent once tried to poison General Washington by feeding him cooked tomatoes.

Incidentally, the received wisdom that poison is a woman's weapon is not borne out in practice. Poison is an equal-opportunity murder weapon.

Arsenic, a metallic element, has been known in its various forms for more than two thousand years. It was used in ancient China as an insecticide and in Rome as a medicine. It is ubiquitous in soil, in water, and in the human body, usually in amounts that are small enough to be considered safe (though there is growing opinion among health experts that no amount of arsenic is truly safe—it is now believed to be a potent carcinogen).

From Roman times on, the custom of eating small quantities of arsenic persisted in various parts of the world. People ate it to improve their complexions as well as their overall health, stamina, and sex drive. They fed it to their horses to increase their endurance and add luster to their coats. Whether arsenic actually conferred any of these benefits is doubtful. A possible exception is arsenic's effects on the complexion. Apparently arsenic ruptures the blood vessels that lie just beneath the skin, imparting a "healthy" ruddiness to the cheeks.

During the earliest years of the Joseon dynasty of Korea (1392–1910), compounds of arsenic and sulfur were used in making *sayak*, a deadly beverage that was used for the official executions of high-ranking court officials and members of the royal family.

Fowler's Solution, an arsenic compound of potassium arsenite and lavender water, was devised by Dr. Thomas Fowler in the 1780s as a remedy for whatever ailed you. The usual dose was twelve drops, three times a week, in a glass of wine or water. It was recommended internally for the treatment of ague, periodic neuralgia, rheumatism, and epilepsy, and as a possible cure for dropsy, syphilis, palsy, nervous tremor, chorea, scrofula, tetanus, and snakebite. Externally, as a paste, it was used for malignant ulceration, lupus, *noli me-tangere* (a cancerous ulceration of soft tissue), fungal infection, and gangrene. It was not recommended for use against fleas, lice, and other bodily pests because the necessary concentrations might prove fatal to the user.

Easy access to arsenic was certainly one of the reasons for its popularity as a "remover of obstacles." Another was that arsenic is odorless and tasteless and thus undetectable in food. As a bonus to poisoners, the symptoms of arsenic poisoning mimic a range of common illnesses, from severe stomach upset to cholera. And even if its presence were discovered, there might be several innocent explanations for its ingestion. Arsenic is widespread, after all—it was once even to be found in wallpaper, green wallpaper especially.

Sometime around 1785 Samuel Hahnemann, whose medical observations led him to develop the practice of homeopathy, discovered that when dissolved in a liquid containing hydrochloric acid and hydrogen sulfide, arsenic would form a yellow precipitate. This became the first practical method for detecting the presence of arsenic in foods or in the bodies of the recently deceased.

In 1787 Johann Daniel Metzger, a professor of medicine at the University of Berlin, discovered an easier way to detect white arsenic in solution. When the suspect material was heated with charcoal, arsenious oxide (white arsenic) would vaporize, if present, and form a shiny black deposit on a porcelain plate.

But supposing the arsenic was in some other form? The Swedish chemist Karl Wilhelm Scheele (1742–1786), an early discoverer of oxygen, chlorine, citric acid, oxalic acid, gallic acid, uric acid, and a host of important chemical processes, also devised a method of producing arsine gas from any arsenic compound that happened to be present in, say, someone's dinner or internal organs. If sulfuric acid or hydrochloric acid is mixed with any liquid containing arsenic, and zinc is then added, the reaction between the zinc and the acid will produce hydrogen, which bubbles through the liquid. The hydrogen will "steal" the arsenic from any arsenic compound present and form arsine gas. If the gas is ignited as a cold porcelain bowl is held to the flame, the shiny black deposit that Metzger discovered will form.

In 1806 Valentine Rose devised a technique that used nitric acid, potassium carbonate, and lime to detect the presence of arsenic in human organs.

In 1832 James Marsh, a chemist at the Royal British Arsenal in Woolwich, was asked by the justice of the peace in nearby Plumstead to take time out from his research into a recoil brake for naval guns to see if he could somehow detect arsenic in a coffee pot. John Bodle, a local farmer, was suspected of poisoning George Bodle, his eighty-year-old grandfather. It was known that John had recently bought some arsenic, supposedly to kill rats, from a pharmacist on Powis Street. This was a week before

George sickened after breakfast with a sudden onset of vomiting, cramps, diarrhea, and weakness in his limbs. George's wife, daughter, granddaughter, and the serving girl, all of whom had breakfasted with him, also developed cramps and stomach pain but recovered. George died.

When Sophia Taylor, the serving girl, told Mr. Slace, the justice of the peace, that John had uncharacteristically volunteered to take the kettle to the well that morning, Slace locked up the coffee pot, ordered an autopsy on Bodle, and sent the coffee and Bodle's intestines to James Marsh, the only qualified chemist around. Marsh, who had no interest in forensic chemistry, had to look up the method for detecting arsenic; he was unaware of the work of either Metzger or Scheele. He used the Hahnemann technique. If arsenic was present, a yellow precipitate of arsenic trisulfide would result. It did. Unfortunately (or fortunately for John Bodle), by the time of the trial the exhibit had deteriorated and was difficult to explain. The British jury found all the messing with acids and gases and yellow precipitates and such to be too much hocus-pocus to be believed, and Bodle was found innocent.

James Marsh found this annoying. He was particularly upset when, sometime later, John Bodle confessed to murdering his grandfather. So Marsh set out to devise a test for arsenic that was both infallible and showy enough to convince a jury. When he read of Scheele's method, he knew this was the place to start. Marsh set out to create an apparatus that would test for arsenic in solution in one step.

First, he took a glass bottle and stoppered it with a two-hole rubber stopper. Then, through one of the holes he inserted a thistle tube—a hollow glass rod with a bulge at one end—to allow liquids to be poured into it. It was inserted thistle side up with the other end extending to the bottom of the bottle. A U-shaped glass tube was then inserted in the other hole, and the bottle partly

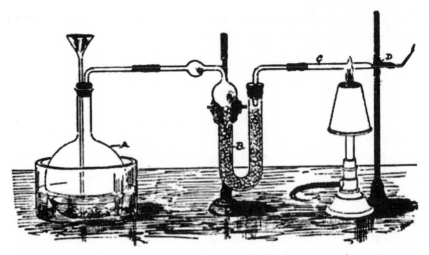

The Marsh test apparatus, designed to test for the presence of arsenic.

filled with zinc and sulfuric acid. This generated hydrogen, which could then be released from the U-shaped tube and ignited. If a mixture containing arsenic was added through the thistle, it would combine with the hydrogen, and the resulting arsine gas would form a dark brown spot on a porcelain bowl. If antimony was present, it would also react with the hydrogen and leave a spot on the bowl that was pitch black and easily differentiated from arsenic.

Several refinements have been made to the Marsh test over the years, and some technicians have found these to be useful—the Berzelius-Marsh method and the Gutzeit test, for example. Today a well-equipped lab within commuting distance of a nuclear irradiating facility will use neutron-activation analysis to determine the percentages of arsenic (and eighty or so other elements) in the sample.

The 1840 trial of Marie-Fortunée Lafarge for the murder of her husband Charles Pouch-Lafarge, known throughout France as *l'affaire Lafarge*, remained a *cause célèbre* for half a century. As a matter of public fascination it was not eclipsed until the 1894 trial for treason of Captain Alfred Dreyfus.

There were those who thought the beautiful Marie must be innocent; there were those who thought her guilty simply because man is defenseless against the insidious machinations of a supposedly loving wife. There were those who believed she was probably guilty but that Charles deserved it—the brute!

Marie Capelle, the daughter of one of Napoleon's favorite officers, a colonel of the Old Guard, was twenty-three when she married Charles Lafarge. She realized immediately that she had made a mistake. He was a coarse, vulgar twenty-eight-year-old with a rat-infested old house in the province of Correze—and, if not a mountain of debt, at least a sizable hill. As soon as they returned from their honeymoon Marie locked herself in her room and slipped a note under the door protesting that she loved another and wanted out of the marriage. That didn't work (and the absent lover was imaginary anyway).

Marie seemed to reconcile herself to her situation, but still the house was filled with rats, and this was too much to bear. She went to the village druggist and bought a large supply of arsenic rat poison. When Charles went to Paris on business for a month, she mailed him a cake to remind him of home. After a few bites he began to throw up. (One of the unexplained facts of the story is that in the letter that accompanied the cake, Marie said she was sending "some little cakes." But what arrived was one large cake. A small point, perhaps, but worthy of consideration according to the "she didn't do it" crowd.)

After Charles returned home from his trip, he quickly sickened again and died. Suspicion soon fell on his young wife when servants said they had seen her stirring a white powder into the dinner. She was arrested for murder and scheduled for trial. At the request of the prosecutors, the contents of Charles's stomach and samples of his uneaten food were examined for arsenic by local doctors. They found arsenic in the food but, curiously, none in his stomach.

At the trial Mme. Lafarge's lawyers questioned the doctors' techniques and results and demanded that the food be tested again. The prosecution in turn requested that Lafarge's body be exhumed and tested for the presence of arsenic in the organs. The body was exhumed, the doctors did the tests, and again no arsenic was found. The food was tested again and, again, more arsenic was found. The French newspapers eagerly followed each new twist in the case and took sides, as did the French public.

With the agreement of both sides, the judge asked Dr. Mathieu Orfila, recognized as the world's greatest expert in toxicology, to come from Paris and test Charles Lafarge's body for traces of arsenic. Using the latest version of the Marsh test, Orfila did a thorough job, examining not only the body but the soil around the body, to ensure that any arsenic he might find had not leached in from the surroundings.

Orfila found slight traces of arsenic in the soil and a great deal more in the body tissues. Mme. Lafarge was found guilty and sentenced to death by guillotine, a sentence that was subsequently commuted to life in prison. She was pardoned after twelve years but lived less than a year after being released. Several people wrote books explaining that she was innocent all along, naming a servant as the true killer. But no one disputed Dr. Orfila's demonstration that there was, indeed, arsenic in Charles Lafarge's body.

Napoleon Bonaparte, emperor of France and conqueror of most of Europe, died in exile on the island of St. Helena on May 5, 1821. He was fifty-two years old. The official cause of death was stomach cancer, but for almost two centuries a rumor has persisted that he was poisoned with gradual doses of arsenic. The only question was, who had done it—the British, or agents of

King Louis XVIII of France, intent on making sure he would never return?

A 1961 analysis of Napoleon's hair found arsenic in it and seemed to confirm the rumor. But in 2008 scientists conducted a study of samples of Napoleon's hair taken from different periods of his life, along with samples from Empress Josephine and from Napoleon's son. The hair was tested at the University of Pavia by a process known as neutron activation, a method that provides extremely accurate results without destroying the sample. The examiners found that the arsenic levels in Napoleon's hair and in the other samples stayed constant, but that the constant level was one hundred times higher than that found in modern samples. In other words, Napoleon was not deliberately poisoned with arsenic, but there was a lot of it in the environment two hundred years ago, and everyone ingested their share.

In 1860 an American explorer of the Arctic regions, Charles Francis Hall (1821–1871), went to Baffin Island in northern Canada in search of some sign of the Arctic expedition of Sir John Franklin, missing since 1845. He returned two years later with artifacts from the Martin Frobisher expedition of 1578, having discovered Frobisher's stone house. His book, *Arctic Researches and Life Among the Esquimaux*, recounts the experience. On his second expedition (1864–1869), Hall did discover traces of the Franklin party.

In 1871 Hall commanded a government-sponsored expedition to the North Pole. His ship, the *Polaris*, reached 82°11' north latitude, a new record, before calling it quits.

While his ship was encased in ice for the winter, Hall collapsed as he drank a cup of coffee. For about a week he seemed to recover, but then he relapsed and died. During his delirium he accused several people, including Emil Bessels, the ship's doctor, of poisoning him.

In 1968 Chauncey C. Loomis, a Dartmouth College professor and biographer of Hall, went to Hall's grave in Greenland and had his body exhumed. Samples of Hall's hair and a single fingernail—all that could be collected from the frozen body— were sent to the Toronto Center of Forensic Sciences for neutron-activation analysis. The center's report found "an intake of considerable amounts of arsenic by C. F. Hall in the last two weeks of his life." The base of the fingernail, grown out during those last weeks, contained 76.7 parts per million of arsenic—three times the amount found in the tip and certainly enough to kill him.

Dr. Emil Bessels, who had a long, contentious relationship with Hall, is the only logical suspect. There the story must end.

Prussian Blue, one of the first synthetic pigments, was probably created by the paint maker Johann Jacob Diesbach sometime around 1706. By 1709 his "Preussisch blau" was sold all over Europe. The first artist known to use it was Pieter van der Werff in his 1709 painting *Entombment of Christ*, now hanging in the picture gallery of Sanssouci Palace, Potsdam.

In about 1783 Karl Wilhelm Scheele added another discovery to his impressive list. Following the lead of the French chemist Pierre Macquer, Scheele was able to derive pure hydrogen cyanide (HCN) from Prussian Blue pigment. It came off as a gas but could be dissolved to some extent in water. This liquid, known as prussic acid, was lethally poisonous. The word "cyanide," from the Greek word for blue, hints at HCN's derivation from Prussian Blue.

Within a dark and dismal cell
In anguish I do lie,
Methinks I hear the solemn knell,
Say Tawell you must die . . .
—English broadside ballad, 1845

On New Year's Day 1845 Sarah Hart of Salt Hill, Buckingham-
shire (now a part of Berkshire), was visited at her home by a
man "dressed like a Quaker." He was observed coming and going from the house twice. That evening several people heard Hart
screaming. They ran to her house where they found her in bed
and suffering from terrible cramps and pain. One of them gave
her water, but foam came to her lips when she tried to drink it.
She later died.

The vicar of Upton-cum-Chumley, who had responded to
her cries, got a description of the absent visitor from a neighbor,
then rushed off to the Slough train station to try to intercept the
man.

Unfortunately the train had just left. But the station had one
of those new telegraph apparatuses, and the stationmaster sent a
message ahead to Paddington Station:

> A murder has just been committed at Salt Hill and the
> suspected murderer was seen to take a first-class ticket to
> London by the train that left Slough at 7.42 pm. He is in the
> garb of a Kwaker [the telegraph instrument had no "Q"]
> with a brown great coat on which reaches his feet. He is in
> the last compartment of the second first-class carriage.

Shortly a reply arrived:

> The up-train has arrived, and a person answering, in every
> respect, the description given by telegraph came out of the
> compartment mentioned. I pointed the man out to Sergeant
> Williams. The man got into a New Road omnibus, and
> Sergeant Williams into the same.

Sergeant Williams followed his suspect to his lodgings in
Scott Yard and arrested him the next morning. His name was
John Tawell.

The autopsy surgeon, H. Montague Champneys, smelled "the odor of prussic acid" when he opened the body. He bottled the contents of the stomach and sent them along with a bottle of beer found at the scene to Mr. Cooper, an analytical chemist and lecturer on medical jurisprudence. Cooper added potassium ferrosulfate to the stomach contents and observed that the mixture turned a deep Prussian Blue, indicating that it contained prussic acid. So did the beer.

When the police found that Sarah Hart had been Tawell's lover for years, that Tawell had recently married a Quaker widow, and further that Tawell had purchased two drams of Scheele's prussic acid from a chemist in London on the day of the murder "to use for his varicose veins," they had all they needed. Tawell was brought to trial.

Tawell's defense counsel, Fitzroy Kelly, attempted an ingenious explanation for the presence of the poison at the scene: apple pips contain prussic acid, and there had been a barrel of apples in Sarah's room. This set off a scramble by medical experts on both sides to see how much prussic acid could be distilled from apple pips. The prosecution found that the amount in fifteen apples would be harmless, while the defense found that the same amount might be toxic.

The jury found Tawell guilty. He was hanged before a crowd of ten thousand on March 28, 1845. Fitzroy Kelly, who went on to have a distinguished career, was thenceforth and forever known to his intimates as Apple Pip Kelly.

John Sadleir (1813–1856) was born into a Tipperary banking family and seemed to have a knack for the business. Starting out in small Irish banks, he worked his way up and eventually moved to London where he was appointed chairman of the London and County Joint Stock Banking Company in 1848. Sadleir then expanded his interests beyond banking. He financed railways in

Italy and France, and started a newspaper, the *Weekly Telegraph*, in Dublin. He bought the Earl of Glengall's estate at Cahir. He owned a stable at Watford from which he hunted with the Gunnersbury hounds. He kept his shareholders happy with a 6 percent dividend when most others were paying 4. He was elected to Parliament. He was appointed a junior lord of the treasury.

In 1853 it all began to unravel. After accusations of fraud, Sadleir was forced to resign his seat in Parliament. He was found to be borrowing money from his own bank, whose declared reserves of £17,000 were found to be nonexistent. He had embezzled some £200,000 from the Tipperary Bank and issued £150,000 worth of phony stock in the Swedish Railway. He could not steal fast enough to keep up.

In February 1856 his cousin received a letter from him in which he confessed to "numberless crimes of a diabolical character," which had caused "ruin and misery and disgrace to thousands—ay, tens of thousands." On Sunday morning, February 16, his body was found on Hampstead Heath. By his side was a silver cream pitcher containing a solution of prussic acid. Dickens used Sadleir as a model for the despicable Mr. Merdle in *Little Dorrit*. James O'Shea, who wrote the 1999 book *Prince of Swindlers, John Sadleir M.P.,* said of him, "He was well-liked and well-trusted and robbed everybody that he could lay his hands on."

The Nux Vomica, or strychnine tree, is an evergreen that grows wild in Southeast Asia. The seeds of its round, green-to-orange fruit are the main sources of the alkaloids strychnine and brucine. Strychnine is intensely bitter and highly poisonous. It was introduced to Europe in the sixteenth century as a rodenticide and is still used on occasion to kill small animal pests. Thirty-two milligrams is a lethal dose for humans, but people have been known to die from as little as five milligrams. It is one of the adulterants used to cut street drugs such as amphetamine and cocaine. Brucine, related to strychnine, is not quite as poisonous and is used in minute quantities in several heart medications.

Dr. William Palmer (1824–1856) had a lust for life and a fondness for horse races, gambling, and women. He also had a tendency to steal when he had the chance and to kill when threatened. When still an intern at Stafford Infirmary he was accused of trying to poison a friend. The charges went nowhere, but the infirmary tightened its procedures for access to the dispensary. Palmer moved back to his hometown of Rugeley in Staffordshire and set himself up as a doctor. In 1847 he married Ann Brookes, and they had their first child the next year. Ann had four other children over the next few years, but they all died in infancy.

People who were close to Palmer had a habit of dying. His mother-in-law passed on, as did at least two people to whom he owed money. In 1854, after he took out a £13,000 insurance policy on his wife Ann, she promptly died. The death certificate read "cholera." Nine months later Eliza Tharme, his housemaid, bore him a son, Alfred. In 1855 Palmer's brother Walter suddenly died, but the insurance company refused to pay Palmer on that one. Palmer, desperate for money, took out an insurance policy on a friend, George Bates, who then conveniently died. A boot boy claimed to have seen Palmer pouring something into Bates's drink just before Bates took sick.

In November 1855 Palmer's friend John Cook won a large sum of money betting on the Shrewsbury handicap stakes. Palmer invited Cook to come with him to Rugeley, and, on November 17, Cook took a room at the Talbot Arms. On that same day Alfred, Palmer's five-month-old son, died. The next day Cook became very ill, and Palmer went to London to collect his winnings for him. At 1 A.M. on November 18 John Cook died.

Cook's stepfather came to Rugeley and demanded an autopsy. Palmer, as a doctor and a friend of the deceased, attended. During the proceedings Palmer was spotted trying to leave the autopsy room with the jar containing Cook's stomach. His bizarre behavior, combined with public sentiment in Rugeley, led to his arrest on suspicion of murder. The bodies of his wife and brother were exhumed and examined for signs of poison, but nothing definite was found.

A special act of Parliament was passed to allow the trial to be moved to London, as public outrage in Rugeley would preclude Palmer's getting a fair trial there. He was tried only for the murder of Cook. The science at that time was not up to the task of finding strychnine in Cook's body, but witnesses described the manner of his death, including severe vomiting, spasms, and powerful convulsions. Medical experts testified that only strychnine poisoning could cause these symptoms. The jury took a little more than an hour and a half to find Palmer guilty.

Palmer was hanged at Stafford Prison on June 14, 1846, with an estimated thirty thousand people in attendance. As he stepped onto the trap door, he allegedly turned to the hangman and asked, "Are you sure it's safe?" There is a persistent story that after Palmer's death the town of Rugeley petitioned to change its name, fearing that it would be forever associated with "Palmer the poisoner." The prime minister agreed, with the proviso that they rename the town after him. His name was Palmerston. The town chose to remain Rugeley.

Castor beans are processed around the world to make castor oil. The mash that is left after the beans are ground contains ricin, the third deadliest poison after plutonium and botulism toxin. The compound was discovered in 1800 by Hermann Stillmark as he was developing methods of extracting the oil from the beans. It is believed that in the 1980s ricin was used as an agent of warfare by Saddam Hussein in Iraq's war with Iran.

A very small amount of ricin—three micrograms—can kill if inhaled, ingested, or injected into the body. Death occurs within thirty-six to forty-eight hours. If the patient manages to hang on for as long as five days, he will probably recover. There is no known antidote for ricin, and a reliable test to detect its presence has only recently been developed by scientists at the Albert Einstein College of Medicine at Yeshiva University. A vaccine is now being licensed by the United States Army.

Georgi Markov, a noted novelist and playwright, defected from Bulgaria in 1969 after writing *The Man Who Was Me*, a play that the Communist government of Bulgaria disapproved of. Markov moved to London and began writing scripts, mostly based on his memoirs, for the BBC and Radio Free Europe. Because of his former access to the upper levels of Bulgarian society, he knew and wrote about many things that the Bulgarian president, Todor Zhivkov, would just as soon not have had aired.

Over the next few years the Bulgarian government made several unsuccessful attempts to assassinate Markov—one in Munich and one on the island of Sardinia. The third attempt, in London, was successful. On September 7, 1978, as Markov waited for a bus on Waterloo Bridge, he felt a stinging sensation in his right thigh as a heavily built man carrying an umbrella bumped into him. The man mumbled "Sorry," and got into a passing taxi. By that evening Markov had developed a high fever and was admitted to St. James Hospital in Balham. Four days later he was dead.

Forensic pathologists did a thorough autopsy of Markov's body at the request of Scotland Yard. They found a tiny pellet that was 90 percent platinum and 10 percent iridium imbedded in his thigh. The pellet measured 1.52 millimeters across—almost exactly the size of a pinhead. Two holes, each .35 millimeters wide, were drilled through the pellet, leaving an X-shaped cavity in the center. Experts from Britain's Defence Science and Technology Laboratory at Porton Downs found that the pellet had been filled with ricin and the holes stopped with a "sugary substance" that was designed to melt at body temperature. Once inside the human body, the substance melted and the ricin was released. At that point nothing could save the victim.

If you happen to have a source of neutrons around the house, say a small nuclear reactor, you have access to one of the best quantitative and qualitative forensic analysis systems yet devised. Neutron-activation analysis involves sticking a sample of just about anything in a small chamber, bombarding it with neutrons, then measuring the gamma rays emitted by the irradiated sample. This method will detect the presence and relative amount of any of about seventy different elements. Best of all, if you're a forensic scientist, the process is nondestructive—the sample is still there and available for other tests or for waving before a jury.

In June 1965 radio personality René Castellani, the "Dizzy Dialer" of station CKNW, Vancouver, sat in a car perched atop the world's largest freestanding electrical sign—a ten-story-high pole on the Bowell Maclean car lot—vowing not to come down until the very last car on that lot had been sold. As he sat there his wife, Esther, was hospitalized with a recurrence of the serious gastric upset that had bothered her for several months. After two weeks, René came down. One month later, in early July, Esther Castellani suffered a severe relapse and returned to the hospital. René hurried to her side from a promotion in a toy store, still wearing his gorilla suit (the zipper had broken). He was just in time to be with her when she died.

Several months after Esther was buried—the cause of death was listed as "a viral infection and heart attack"—a woman approached the authorities and claimed that Esther had been poisoned. She would tell all, she said, if she were given immunity. She was insistent enough and her story provocative enough that Esther's body was exhumed and tested for arsenic. And it was found—a lot of it. Possible sources of contamination—the ground she was buried in and the embalming fluid—were eliminated, and it became clear that Esther had been deliberately poisoned. The informant claimed that she and René were lovers, and that René had poisoned Esther for the insurance money.

René Castellani was brought to trial for the murder of his wife. That she had been poisoned with arsenic was not in dispute. But the defense suggested that the lover had done it to eliminate her rival, and the prosecution would have to prove otherwise.

Samples of the victim's hair were given to forensic chemist Norm Erickson who cut them into precise lengths corresponding to the amount that hair grows in one week, two weeks, and a month. When the lengths were subjected to neutron-activation analysis, a very interesting fact appeared: all the hair showed a concentration of arsenic. Esther had been fed small doses of it regularly. But during one two-week period the hair was nearly arsenic-free. During this period she had not ingested any of the poison because her husband René had been in a car atop the world's largest freestanding electrical sign.

René Castellani was convicted of the murder of his wife and sentenced to twenty-five years in prison.

Newer and more subtle poisons are now available. Some of the drugs used as anesthetics will kill quickly with no outward sign, and will disappear from the body completely in three days or less. If the victim suffered from a serious ailment, suspicion would not be aroused. The greater worry, if you're inclined to worry about such things, is the lack of suspicion among medical professionals. After all, most people aren't poisoned. Or so we like to think.

11 : Blood Will Tell

THE IMPORTANCE OF BLOOD was recognized centuries before its biological function was understood. "The blood is the life," as Count Dracula told Jonathan Harker in Bram Stoker's novel; if your "life's blood" ebbs out, you will die. Patriots are "red-blooded Americans." The aristocracy are "blue bloods." An evil person has "bad blood" and an insane person "tainted blood." Blood, after all, is thicker than water.

Blood is the one thing investigators are almost certain to find in abundance at crime scenes involving physical violence. The average adult body contains a lot of it, about five liters (ten pints). Most injuries release at least a bit of it. It remains liquid for quite a while after leaving the body, and whatever it lands upon it stains. A perpetrator is quite likely to take a bit of it away with him. As the criminalist Harry Söderman put it in his autobiography *Policeman's Lot*, "Blood is a liquid which, in a crime of violence, seems to have an uncanny capacity to hide itself, only to reappear at a fitting moment and testify against the murderer. Blood will creep under the tiles of a floor, into the cracks of boards and the grain of wood, under fingernails; blood will linger in the water trap of a basin where the killer has washed his hands. It will cling beneath the top of a table where fingers have been thoughtlessly wiped. It splashes on clothing and into hair."

Imagine that the police have a suspect in custody and that there are suspicious brown stains on his trouser legs.

"It's paint," he claims.

Is it blood?

"Okay, it's blood. I was hunting and shot a pheasant."

Is it human blood?

"Okay, it's human blood. I had a nosebleed."

Is it his blood, or is it the victim's?

It took more than a hundred years of forensic research to find the answers to these questions.

Bloodstains are not stable but rather change with the passage of time. They are influenced by the effects of temperature, moisture, and sunlight. They are not always red or brown but can be altered by environmental factors in such a way that they may appear grey, tan, yellow, or even green. Fresh blood can be identified under the microscope—the erythrocytes (red blood cells) and leucocytes (white blood cells) show up clearly. But as blood dries, the two cell types clump together and cannot be differentiated. If the stain is not too old or too small, the red cells may be coaxed out again by washing the stain in a solution of caustic potash and alcohol. Sometimes this is successful, and sometimes not.

In 1853 the Polish physician Ludwig Teichmann developed a test for blood that was complex but effective when it worked. He found that if a bit of the suspected blood was dissolved in a mixture of saltwater and glacial acetic acid, and then warmed, crystals of hematin would form as a positive reaction to the presence of blood. The crystals are a sure sign that the substance is blood. Sometimes, if rust is present in the blood, the crystals may not form. But if they *do* form, the substance is blood.

Ten years later a German chemist, Christian Friedrich Schönbein, the discoverer of ozone and the inventor of the fuel cell, found that hydrogen peroxide would foam in the presence of blood. Even tiny amounts of blood would set off this reaction. But so would tiny amounts of semen, saliva, rust, and certain shoe polishes. So Schönbein's test is a quick and good one for elimination purposes—if it does not foam, it is not blood. But if it does foam, it may not be blood and you must test further.

The reaction is set off by the blood enzyme catalase, which turns the hydrogen peroxide into water and oxygen. The oxygen is given off as a gas and thus causes the foaming. Catalase is the blood component that pulls oxygen out of the air in the lungs and delivers it to all the cells of the body. Its oxygen-grabbing ability is strong.

A quick test, though not a positive one, for blood in a dried stain uses tincture of guaiac, the resin of any of the six trees or shrubs of the genus *Guaiacum*, a subtropical evergreen with several varieties that are grown today as ornamentals in Florida and California. Guaiac is still used in a medical test that screens for minute traces of blood in feces.

The procedure in this case is to cut out the stain, lay it on a clean sheet of glass, moisten it with distilled water, and let it sit for a time. A technician then covers it with a sheet of filter paper cut to size, presses the paper down firmly with a glass rod, and waits a while longer. Then the technician takes up the filter paper and tests it with a few drops of an equal mix of tincture of guaiac and turpentine. The presence of blood will color the filter paper blue, as will a few other substances. As with the test for crystals of hematin, this is a useful test for the elimination of a suspected stain. But if the paper does turn blue, you will need additional samples of the stain for use in further testing.

By the end of the nineteenth century a wide array of tests for blood had been developed. The most reliable and certain of these was the spectroscopic examination of the stain. But very few forensic labs had access to a spectroscope, and there were still no tests that could differentiate human blood from animal blood.

> The importance of a means of recognizing human blood from that of other mammals is very apparent, but is a problem that does not appear to be well understood. Some years ago Mr. Barreul, a chemist of Paris, became so skilled in the recognition of animal blood by the agency of chemical action in connection with his unusual acuteness of the sense of smell, that his testimony was taken in the Paris courts as positive evidence. We believe the process consisted in the addition of sulphuric acid to the blood, and the test consisted in the odor evolved during the heat resulting from the mixture. We don't recollect whether this ability to detect included dried blood or not, but remember that it detected unmixed blood with great certainty, the blood of each animal evolving a peculiar odor.
>
> —*American Journal of Pharmacy*, July 16, 1866

According to M. Barreul, when boiled in sulfuric acid cow's blood smelled like the inside of a barn, sheep's blood smelled like grass, and human blood smelled like sweat. Since few others could duplicate the feats of his remarkable nose, this investigative technique was soon abandoned.

One accurate, if slightly fallible, method of telling human blood from that of other mammals could be employed by chemists other than M. Barreul. If the blood was fresh and had not yet clotted, it could be differentiated microscopically by the size of its red blood cells. Birds, fish, and reptiles all have oval, nucleated (containing a nucleus) red blood cells. All mammals have circular red blood cells without nuclei (except for camels and llamas, which have oval cells, but still no nuclei). From mammal to mammal, these cells vary in size. Human erythrocytes have a diameter of 0.0003 inch (three ten-thousandths of an inch), cow cells average 0.00025, and goats 0.00015. It takes a good microscope and a good eye to read a difference of less than one ten-thousandth of an inch. Further, when blood dries the red cells are deformed, and it becomes difficult to restore them in a way that allows accurate determination of their size. A better test for human blood was needed.

The February 7, 1901, issue of the *Deutsche medizinische Wochenschrift* (*German Medicine Weekly*) contained an announcement by Paul Uhlenhuth of the Institute of Hygiene at the University of Greifswald that he had devised a test that would distinguish human blood from any other, no matter how small the trace. Uhlenhuth's claim was greeted with profound skepticism. An assistant professor at a small, unimportant university had succeeded where some of the brightest and most respected men in medicine had failed? Unlikely.

But Uhlenhuth, in the best tradition of science, had stood on the shoulders of those who had come before him and peered in a slightly different direction. In 1890 Emil Adolf Behring and his co-workers at the Institute of Hygiene in Berlin had developed a method for immunizing people against diphtheria and tetanus. By injecting guinea pigs with a filtrate of the diphtheria culture from which the actual bacilli had been removed (a substance they called a toxin), they were able to induce a reaction in the blood of the guinea pig that would neutralize the diphtheria. And the guinea pig's blood serum (the watery liquid remaining when all the cells and detritus are removed) could confer this protection to other guinea pigs, rabbits, and humans. They called this blood serum an antitoxin.

This breakthrough prompted the founding of serology, an entirely new branch of medicine. In the same year that Uhlenhuth announced his discovery, Behring won the Nobel Prize for his groundbreaking work.

The term "serology" originally meant only the study of serum, but so much more has been discovered about blood over the years that the term now includes all the various laboratory tests that identify antibodies, antigens, and the numerous other substances found in blood.

Uhlenhuth found that if he injected protein from a chicken egg into a rabbit and then, after a time, took some of the rabbit's blood and created a serum, he could mix the serum in a test tube with chicken egg white and a cloudy precipitate would form and drift to the bottom of the tube. But it worked only for chicken egg white; when the whites of pigeon, gull, or turkey eggs were mixed with the serum, nothing happened. The injection of chicken egg protein into the serum had somehow primed it to differentiate chicken protein from that of any other bird. He quickly learned that the same thing happened if he used pigeon egg protein. When injected with pigeon egg protein, the rabbit's serum would precipitate out the protein of no other bird.

When Uhlenhuth tried developing a serum with chicken blood instead of egg white, a flaky protein precipitated out and dropped to the bottom. Uhlenhuth had discovered how to create a serum that would react with the blood of one animal and one animal only. (There were some closely related animals whose blood was too much alike for the test to work—horses and donkeys, for example, and humans, gorillas, and chimpanzees.) His test succeeded even with small amounts of very old, dried blood as long as the blood was first dissolved in saltwater.

Uhlenhuth soon had serums made for his so-called precipitin test that would be specific for every animal imaginable. And he quickly developed safeguards for the test. After faulty results were obtained from serum made in another lab, he standardized the serums and insisted that his own institute, along with the Robert Koch Institute in Berlin, be the only official sources. He also strongly suggested that before each test of an unknown substance, a control test be conducted against a known sample. When some confounding results occurred due to the background material of the stain—tree bark, for example—Uhlenhuth added the suggestion that the underlying material be tested first in order to eliminate the possibility of a false positive. With these safeguards, the Uhlenhuth precipitin test worked infallibly every time.

On June 11, 1904, most of the body of a young girl was found tied up in a not-too-neat bundle of wrapping paper. The corpse floated in the Spree River between the Alsen Bridge and the Weidendamm Bridge, almost opposite the Reichstag in Berlin. The child's torso, minus the head, arms, and legs, was clad in bloomers and a red woolen petticoat. Within an hour the body was identified as that of nine-year-old Lucie Berlin, youngest daughter of Friedrich Berlin, a cigar maker. She had been missing from her home at 130 Ackerstrasse for two days. The police immediately offered a reward of a thousand marks and ran the following notice in the newspapers and on broadsides posted around the city:

At 7:45 A.M. today the torso of Lucie Berlin, born in Berlin on 8 January, 1895, was washed up in front of the building at 26 Schiffbauerdamm. Head, arms, and legs of the corpse were severed with some sharp instrument. The girl was last seen playing in the yard of her home at 130 Ackerstrasse from noon to around one o'clock on the ninth of this month, and has been missing since then. She was tall for her age, was wearing a russet-brown wool dress, a black pinafore, white stockings, brown bloomers, buttoned shoes. She had a rectangular gold locket around her neck on a black velvet ribbon. Evidently the girl was the victim of an indecent assault. Persons who can offer information regarding her whereabouts from 9 June to 11 June, or who encountered

her, are requested to communicate with the police at
headquarters on Alexanderplatz or at any precinct.

—Police Chief von Borries

Berlin did not yet have an official medical examiner in 1904,
so the child's autopsy was performed by Professor Fritz Strass-
mann, head of the Academy for Instruction in Public Health, and
by his associate Dr. Schulz. The child had probably been strangled, they reported. Her vagina had been torn by rough fingers,
but probably after she was dead or near death through strangulation. She had died about an hour after eating her last meal of
"pork, potatoes and cucumber salad."

Murders were not common in this turn-of-the-century city
of four million—there had been only thirty-eight in all of 1904.
So the death and dismemberment of a young girl was front-page
news. The headline in the *Berliner Tageblatt* read "Entdeckung
eines Lustmordes" ("Discovery of a Sex Murder"). The *Berliner
Morgenpost* published an extra edition and hung posters all over
the city. Berliners were on the alert for "brat chasers," as those
adults who were overly fond of children were called. Several had
to be rescued by the police from spontaneously formed mobs.

Police suspicion soon settled on Theodor Berger, a thirty-
five-year-old junk dealer who was visiting his girlfriend, Johanna
Liebetruth, a prostitute who lived down the hall from the murdered girl. In fact Berger was not actually visiting her; he was
waiting in her apartment to welcome her home from a three-day
stay at the women's prison on Barnimstrasse. Berger had actually
lived with her in the apartment ever since she had moved in six
months earlier. The address on his identification document was
a domicile occupied by another fellow. And Berger was not actually a junk dealer, but he hoped to become one some day. Aside
from living off the earnings of Johanna, it was unclear how he
made his living.

Several things pointed to Berger's involvement. He knew the girl—she called him "Uncle." Anna Müller, an eighty-year-old woman had seen Lucie on the stairs the day she disappeared and had noticed Berger ogling the girl as she passed his door. Frau Marowski, who occupied the apartment above Johanna Liebetruth's, had heard a child scream "No!" somewhere below at around one in the afternoon. At 1:30 Herr Nölte and his wife, who lived below Liebetruth, had heard a loud series of thumps from above—"as though someone had fallen out of bed," he told the detectives.

And then there was the interesting fact that on the very day Johanna Liebetruth got out of prison, Theodor Berger had suddenly agreed to marry her, something she had been after him to do for eighteen years. Was marriage perhaps the price of her keeping her mouth shut?

Two days after the torso was found, some boys fished a child's head and two arms out of the Charlottenburg ship canal. The body parts were wrapped in a copy of the *Berliner Morgenpost* dated June 9, the day Lucie went missing. Lucie's right leg was soon found in another canal and her left leg in the Spree. Now most of Lucie's body was assembled. Someone had gone around distributing parts of Lucie in bodies of water all over Berlin. But was it Berger? The evidence against him—the best if not the only suspect—was circumstantial and far from convincing. The police searched the Liebetruth apartment for signs of blood; they pressed large pieces of blotting paper soaked in hydrogen peroxide or tincture of guaiac everywhere. They got a few reactions that might have been blood, but nothing to indicate the large amount of blood that must have flowed from poor Lucie's dissected body. And there was none on what were usually the most incriminating spots—the sink drain, the rug, the kitchen knives, and articles of clothing.

It occurred to the police to wonder what the dismembered body had been carried in. Surely the murderer had not wandered about Berlin with newspaper-wrapped body parts under his arm. What might he have used? Johanna Liebetruth told an investigator that when she returned from jail a large wicker suitcase was missing from her rooms. At first Berger had denied knowing anything about it, but when she pressed him he had shamefacedly admitted that he had picked up a girl when Johanna was away and had paid for her services with the wicker suitcase. He didn't know the girl's name and had not seen her before or since. Even Johanna was not sure she believed Berger's story.

The police let it be known that they were looking for a large wicker suitcase that was probably in one of the bodies of water around the city. On Sunday, June 26, twenty five days after Lucie's disappearance, a bargeman named Wilhelm Klunter reported finding it. He had actually fished it out of the water some days before, but because he never read a newspaper he was unaware that the wicker suitcase was wanted until told about it by his aunt.

Liebetruth examined the suitcase and said it was hers, pointing out several identifying features. And there were dried, dark red spots on the side of the suitcase that might well be blood. But was there any way to tie the suitcase to Lucie Berlin's body? Proving that the spots consisted of human blood would be a large step in that direction.

Dr. Schulz had practiced the Uhlenhuth precipitin test, and here was an opportunity to put his new skills to use. He scraped one of the stains off the wicker and dissolved it in distilled water. Following Uhlenhuth's instructions to the letter, he also made solutions of scrapings from an unbloodied bit of the wicker suitcase as well as a known solution of human blood.

When he added a measured amount of diluted serum containing antibodies for human blood to these solutions, it took only seventy seconds for clouds of precipitate to form in the tubes containing the wicker stain and the known blood sample. Only the plain wicker sample stayed clear.

The stains on the suitcase were from human blood.

On December 23, 1904, after a twelve-day trial, Theodor Berger was found guilty of the murder of Lucie Berlin.

Several years before Berger's trial another important step in blood identification had been taken, a step that for the first time allowed for safe blood transfusions. It saved the lives of countless surgery patients and trauma victims. On November 14, 1901, Dr. Karl Landsteiner, an assistant at the Department of Pathological Anatomy at the University of Vienna, published a modest paper in the *Wiener klinische Wochenschrift* (*Vienna Clinical Weekly*) entitled "On Agglutination Phenomena of Normal Human Blood."

Dr. Landsteiner had discovered that some human blood did not mix well with other human blood, that a serum of one person's blood might make another person's blood agglutinate, that is, clot up. Landsteiner had no idea why this happened, but he was determined to find out. He went around his department taking blood from everyone who would donate and tested each sample against the others.

It took Landsteiner a while, but he finally determined that there were four blood types, which he eventually named A, B, AB, and O. The letters indicate the different types of a protein called an antigen that is found on the surface of red blood cells. Type A blood cells have the A antigen, type B have the B antigen, type AB have both, and type O have neither. And because blood serum contains specific antibodies that react with (or not) the different types of antigens in the blood cells, this means that:

Type A antibodies will agglutinate Type B blood.

Type B antibodies will agglutinate Type A blood.

Either serum will agglutinate type AB blood.

Neither serum will agglutinate type O blood.

What this means practically is that if you know what blood type a person has, you know what types of blood you may safely transfuse into him. Type O blood may be given to anyone; type A blood may be given to people with type A or AB blood; type B may be given to people with type B or AB blood; type AB blood may be given safely only to people who have type AB blood.

An additional complication, discovered by Dr. Landsteiner and Dr. Alexander Wiener forty years later, is the Rh blood factor, another protein found on the surface of red blood cells. It is named after the Rhesus monkeys used in early tests. Roughly 85 percent of the population are Rh positive (Rh+), that is, they have the Rh factor in their blood; and 15 percent are Rh negative (Rh–).

This has practical consequences in pregnancy. When an Rh negative mother carries an Rh positive baby—which happens if the baby inherits the father's Rh positive blood type—the mother's immune system may develop antibodies against the blood of her own baby. Usually this does not affect the first pregnancy. But if during a second pregnancy the fetus is again Rh positive, the antibodies developed during the first pregnancy can mistake the fetus for a foreign substance, enter it through the umbilical cord, and attack the baby's blood. This may cause severe jaundice and anemia in the newborn child and may even result in the child's death. Now that the condition is understood, an Rh negative mother may take several precautions that will protect her child.

For forensic serologists, the Rh factor provides one more handle with which to identify an evidential blood stain. When added to the O, A, B, and AB blood factors, it is useful for elimination— that is, if none of the blood factors matches the suspect's, it is not his blood. But it does help narrow down blood types into small enough groupings to tie a given suspect to a crime. If a blood sample is type O, Rh positive, roughly forty people out of every one hundred in the United States match it. If it is AB negative, the rarest blood type, only three people in every ten thousand match it. Still, this means that in the city of Los Angeles alone there might be three thousand matches.

According to the American Red Cross, the breakdown in frequency of blood types in the United States is as follows:

O+   40 percent
O–   7 percent
A+   32 percent
A–   5 percent
B+   11 percent
B–   1.5 percent
AB+ 3 percent
AB– .05 percent

In the past several years a variety of secondary blood differences have been found, secondary because they do not affect the safety of blood transfusions and are of interest only to geneticists and forensic serologists. Some of these secondary factors are the M, N, MN, and P factors, which in certain combinations have been linked experimentally to a tendency to hypertension.

There are other factors that are differentiated by electrophoresis, a technique in which a cotton thread is saturated with the sample blood and then embedded in a thin gel. When a direct current is passed through the gel, the various components in the blood move down a glass plate at slightly different speeds. When the current is removed, the plate is stained and the blood factors now appear as dark bands spread out along the plate. Unlike ABO typing, which has been used successfully on ancient mummies, or DNA, which can persist for eons, this technique works best with comparatively fresh blood, and the differences will fade away after only a few weeks.

Other secondary blood factors are:
Adenosine Deaminase (ADA)
Adenylate Dinase (AK)
Erythrocyte Acid Phosphatase (EAP)
Esterase D (EsD)

Glyoxalase I (GLOI)
Group-Specific Component (Gc)
Haptoglobin (Hp)
Hemoglobin (Hb)
Phosphoglucomutase (PGN)
Transferrin (Tf)

If tests for these factors are available, law enforcement can narrow its range of suspects to well under 1 percent of the population. This comes nowhere near the one-in-ten-million-or-better possibilities of DNA, but these tests may be performed more quickly than DNA screening. They are very effective in eliminating suspects.

In addition to deductions about the source and composition of the blood left behind at a crime scene, observations of the blood's position at the scene may also yield a great deal of information. Blood that has been dropped, dripped, spattered, sprayed, or splashed in a particular direction and at a particular height tells the story of the crime to a trained analyst. Given the right spatters, it is possible to determine where the victim and the killer stood, which hand the killer used, whether or not the victim fought back, whether anyone else was in the area, and other details that may eventually prove relevant.

In the early morning hours of July 4, 1954, the Bay Village, Ohio, police were called to the home of Dr. Samuel Sheppard, a wealthy and popular osteopathic surgeon. Downstairs they found a battered and barely coherent Dr. Sheppard being tended to by the mayor of Bay Village, Sheppard's next-door neighbor. Upstairs in a bedroom they found the body of Marilyn Sheppard, Dr. Sheppard's wife. She had been savagely beaten to death.

Bay Village, an exclusive suburb of Cleveland, had never before experienced a crime even remotely like this one. The murder quickly became the subject of intense interest and even passion.

Sheppard said he had been sleeping on the living room couch when he was awakened by the sound of Marilyn's screams coming from the bedroom. He rushed upstairs but was knocked down— and out—by a "white figure" on the staircase. When he came to, he groggily went into the bedroom and saw that Marilyn was dead. As Sheppard went downstairs, he saw the so-called white figure leave the house through the back door. He gave chase, then grappled with a "bushy-haired form" before he again lost consciousness. The next thing he remembered was waking up sometime later on the beach (his house faced Lake Erie). He rushed home and ran upstairs to find that Marilyn was indeed dead in the bedroom. He then went back downstairs, called his next-door neighbor, Mayor John Spencer Houk, and then collapsed on the living room couch. Houk and his wife hurriedly dressed—it was then around 5 A.M.—and rushed over. Houk tried to rouse the dazed and confused Sheppard while his wife went upstairs where she found Marilyn Sheppard's battered body.

In a room drenched with blood, Cuyahoga County Coroner Sam Gerber found Marilyn lying face up on the bed. There were more than twenty deep cuts to her face and body. Under Marilyn's body he found a pillowcase with a bloodstain on it that seemed to him to have the outline of a surgical instrument. The specific instrument could not be identified, but the term "surgical instrument" seems to have hypnotized the police. They were already settling on "Dr. Sam," as he was known locally, as the murderer of his wife Marilyn.

When Sheppard went to the sheriff's station for a voluntary interview, six days after the murder, the police were finished settling. "Did you ever," they sprang at him, "have an affair with a Sue Hayes?"

"We're just good friends," he replied.

On July 16 Louis B. Seltzer, editor of the *Cleveland Press*, printed an editorial headed "The Finger of Suspicion," which deplored "the tragic mishandling of the Sheppard murder investigation" and suggested that Sam Sheppard hadn't been arrested yet only because of the incompetence of the Bay Village police and the influence of the Sheppard family. On July 20 Seltzer placed an editorial on the front page, under the five-column headline "Getting Away with Murder," which shamed the Bay Village Council into asking the Cleveland police to take over investigation of the case.

At the coroner's inquest, a friend of the family, Nancy Ahern, was pressured to admit that Marilyn Sheppard had told her of Sam's affair with Susan Hayes and had said she was afraid Sam would ask for a divorce.

On August 16 a grand jury met to consider the evidence against Sheppard. The next day he was arrested and charged with the murder of his wife.

The trial began soon after the jurors took a tour of the Sheppard house, where they saw all the blood in the bedroom. Dr. Lester Adelson testified for two days that Marilyn Sheppard had been beaten to death. Patrolman Fred Drenkan testified that he had heard Sheppard's story about fighting with a bushy-haired man but did not believe it. Coroner Sam Gerber told the jury about the bloody pillowcase, declaring, "In this bloodstain I could make out the impression of a surgical instrument." The pillowcase was handed over to the jurors to examine.

A fingerprint expert testified that he found Sam Sheppard's left thumbprint on the headboard of Marilyn's bed. He agreed on cross-examination that it could have been there since well before the murder. After a passel of expert witnesses, the prosecution put Susan Hayes on the stand to testify about her affair with Sheppard, and then rested.

The defense had Sheppard's brother, Dr. Steven Sheppard, testify to the extent and seriousness of Sam's own wounds—but since Steven was Sam's brother, he may not have convinced the jury. Sam Sheppard took the stand and told about the murder night's events, but his wooden delivery made him a bad witness. In cross-examination the prosecutor concentrated on Sheppard's extra-marital affair and occasional dalliances, and finished by asking, "Isn't it a fact that you beat your wife to death?"

"No, sir," Sheppard replied.

It took the jury eighteen ballots to find Sheppard guilty. After his conviction, the defense was for the first time allowed to examine the crime scene. Bill Corrigan, Sheppard's defense attorney, had not pressed to get the keys to the house during the trial, not realizing the importance of the blood evidence. When the prosecution refused to hand them over, he had not protested. But now that the trial was over he decided to find out if an expert could tell what had happened that night by examining the bloodstains. Corrigan made a long-distance call to the University of California at Berkeley and spoke to Dr. Paul Leland Kirk.

Dr. Kirk arrived in Bay Village on January 22, 1955, eight months after the crime. He examined the prosecution's exhibits: the pillowcase with the "surgical instrument" bloodstain; Dr. Sheppard's trousers; and the wristwatch Sheppard had been wearing, which had moisture inside it and a bloodstain on it.

Even eight months after the crime Dr. Kirk was able to reach a closely reasoned and intensely detailed conclusion, one that was vastly different from the one reached by the police, the prosecuting attorney, and the jury.

Dr. Kirk recreated the crime from the blood spatters in the bedroom. The blood spots on two of the walls were caused by the direct battering of Marilyn's head, and those on a third wall were spatter marks from a swinging weapon. A heavy flashlight, or something similar, was probably the weapon used. There was no indication of any "surgical instrument"—the stain on the pillowcase had been caused by its being folded while wet with blood.

There was a blood void—that is, an absence of blood—on one side of the bed where the killer must have stood. Considering the amount of blood everywhere else, the killer would have carried a lot of blood away with him. Sheppard showed no such massive spatter on himself or his clothes. Dr. Kirk also determined that the killer had held the weapon in his left hand. Sheppard was right-handed.

Despite this and other indications of Sheppard's probable innocence, the defense's request that Sheppard be granted a new trial based on new evidence was denied. All this evidence, the court decided, should have been presented at the first trial. The fact that the defense was locked out of the murder house until after the first trial the Ohio appeals court found irrelevant.

Sheppard spent nine years in prison before finally being granted a new trial. By this time his attorney had died and attorney F. Lee Bailey, later to be famous, had taken over as his lead counsel. At his second trial in 1966, Dr. Sam Sheppard was found not guilty. A strong suspect for the murder came to light, but he died before any action was taken.

As Dr. Kirk said, "No other type of investigation of blood will yield so much useful information as an analysis of the blood distribution patterns." But it takes training and experience to analyze blood spatter properly. The tendency to see in the patterns what the investigator hopes or expects to see must be rigorously resisted.

12 : DNA Will Tell More

Much of forensic science may be simply
applied commonsense, but one does have
to collect some scientific data to apply it
to. It is this data-collection which is getting
more and more technical, even esoteric. We
started 35 years ago with test tubes; now
we use gas chromatography and electron
beams. A modern forensic laboratory
is full of mysterious boxes of electronic
gadgetry automatically printing out slips
of paper and drawing wiggly lines on
charts. Most of these mysterious machines
were not invented, or at least not available
commercially, until after World War II.
Now they are spreading like dry rot and
keep on proving themselves indispensable
everywhere. The forensic scientist trying to
keep up with their proliferation is like the
Red Queen—he has to run as hard as he can
to keep in the same place.
—Dr. Hamish Walls, *Expert Witness*, 1972

IF DR. WALLS, the former director of Scotland Yard's Metropolitan Police Laboratory, could see what today's crime lab looks like and how rapidly new instruments and techniques are being introduced, he would wonder that anyone could keep up. These many new techniques include but are by no means limited to scanning electron microscopy, inductively coupled plasma-optical emission spectroscopy, imaging laser-ablation mass spectroscopy, restriction fragment length polymorphism testing, thin-layer chromatography, and atomic absorption chromatography.

Before becoming a standard component of forensic analysis, each of these technical advances had gone through a slow process of acceptance, first by the scientific community and then by the legal community.

Perhaps no single technological advance has had a greater impact on law enforcement than the discovery of DNA and the subsequent development of DNA profiling. Now the ability to identify suspects and to establish their presence at a crime scene has increased exponentially. But are these technologies as infallible as they are reputed to be?

Let's begin with a brief primer. DNA is the stuff that people are made from. In fact every living thing comes from a blueprint furnished by the *deoxyribonucleic acid* molecule. We were a long time in figuring this out.

Many centuries ago we concluded that some traits of appearance and even character are inherited. This is true in the general sense—rabbits give birth only to other rabbits—and in the more specific sense—a child often has the same eye color, hair color, height, and disposition as his parent. Then, in 1866, Gregor Mendel (1822–1884), an Augustinian priest, published the results of a series of experiments with pea plants (*Pisum sativum*) in which he advanced the idea of dominant and recessive genes. These "factors," as he called them, control certain traits—wrinkly skin, for example, if you are a pea plant, eye color if you are human.

A couple of years after Mendel's paper, Friedrich Miescher, a Swiss physician, discovered a substance he called "nuclein" in the nuclei of cells. We now call it nucleic acid, the NA in DNA and RNA.

The discovery of the actual function of Miescher's nuclein and of DNA is a complex story woven of brilliant science, chance discovery, intense rivalry, and two Nobel Prizes. Linus Pauling was awarded one in 1954 for his work on the chemical bonds and structures of molecules and crystals; in 1953 he proposed that the DNA molecule was shaped like a triple helix—three spiral staircases entwined. He was only one helix off.

In 1953 James Watson and Francis Crick got it right, aided by an x-ray diffraction image known as Photograph 51 taken by Rosalind Franklin. The actual shape of the DNA molecule, they discovered, was a double helix. They won the Nobel Prize in 1962.

DNA has been described variously as the "blueprint" of life, the "chemical building block," and the "book of life," in an attempt to encompass its fundamental importance to life on Earth. Each strand of the DNA double helix is made up of multiples of four organic compounds called "bases," arranged in a long string: adenine (A), guanine (G), cytosine (C), and thymine (T). In primitive organisms the two strands, effectively mirror images of each other, separate and then rebuild their other halves, thus creating two double helixes of DNA where only one existed before. Thus a cell can divide and become two identical cells.

In sexual reproduction the male and female each contributes half of the genetic material to create a new and unique individual. And since each strand is picked more or less at random, no two beings who are not identical twins have exactly the same genetic makeup.

In 1984 Alec Jeffreys, a geneticist at the University of Leicester, was comparing DNA patterns of the members of a family, in search of evidence of the genes that caused hereditary diseases. Jeffreys had found that the human DNA molecule contains many

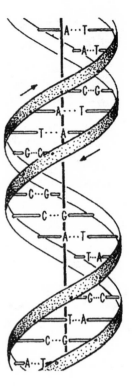

The DNA double helix, discovered by Watson and
Crick in 1953.

places where the AG and TC bonds appear in repeating sequences
of the same pattern, sometimes called a "stutter." A simple stutter would look like this:

    AAGGTC
    TTCCAG

Repeated six times, it appears as follows:

    AAGGTCAAGGTCAAGGTCAAGGTCAAGGTCAAGGTC
    TTCCAGTTCCAGTTCCAGTTCCAGTTCCAGTTCCAG

Jeffreys found that these repeating patterns, known as "mini-
satellites," formed sequences of different lengths in different
people's DNA. At the site where a particular repeating pattern
occurred, some people might have twenty repeats of the pattern,

others fifteen, others thirty-one, and so on. Jeffreys used a "restriction enzyme," a chemical that cuts the DNA molecule into pieces at the same place in each molecule, to split those pieces (called "restriction fragments") apart. Then, using a series of complex techniques, he created a radioactive copy of each DNA fragment. This irradiated copy, called a "probe," left an image, a dark spot called a "band," on x-ray film. And since every person carries two strands of DNA, one from each parent, each probe produced an image of two bands.

At work one morning, Jeffreys, in what he called a "eureka moment," suddenly realized that because of the vast number of possible patterns of repeating sequences, no two people were likely to have the same DNA pattern—that the DNA pattern for each human being would be as individual as his or her fingerprints. Jeffreys's realization would soon revolutionize the science of forensic identification.

In 1985 Jeffreys's lab was asked to help determine whether a boy whose mother wished to bring him from Ghana to live with her in England was really her child. If he was, he could legally immigrate; if not, he would be deported. The DNA patterning showed that the boy shared all the DNA fragments that Jeffreys measured with either his father or the woman who claimed to be his mother. The test proved "beyond any reasonable doubt" that she was telling the truth; the boy was her son.

Soon afterward, Jeffreys's new techniques received even more publicity when they were used to establish the paternities of children in several high-profile divorce cases.

Then, in 1987 the police in the village of Narborough, near Leicestershire, turned to Jeffreys for help. A serial rapist-murderer was loose in the area. His first killing had occurred four years earlier, when the body of fifteen-year-old Lynda Mann had been found on the grounds of the Carlton Hayes Psychiatric

Hospital near the town. Now a second fifteen-year-old, Dawn Ashworth, had been found raped and strangled in the village of Enderby. The most sophisticated blood tests available at the time had established that both girls had been raped by a secretor—a man whose blood type showed up in his semen—and that the blood type was PGM1 + Group A. This eliminated 90 percent of the men in England, but it was not enough to convict the man they had in custody, the person they believed to be guilty of the crimes.

The suspect, seventeen-year-old Richard Buckland, was a porter at the Carlton Hayes Hospital. The learning-disabled young man allegedly knew details of Dawn Ashworth's murder, had made several confused admissions about it, and had then retracted them. But he refused to admit or discuss anything about the first murder, that of Lynda Mann, who had been found on the hospital grounds.

Serologic testing of his blood revealed that he had blood type PGM1 + Group A, the same as that of the semen found in the victims' bodies. With a DNA match, the police felt certain they could clinch the case against him.

Jeffreys's lab ran DNA profiles of the semen from the victims' bodies and of the blood of the suspect. The DNA typing showed that both victims had been raped by the same man, but that the rapist's DNA did not match Buckland's. The first DNA test ever done in a criminal case thus resulted in the exoneration of the accused man.

The police were now back to square one. They decided to try a sweep of local men to see if a DNA match could be found. From the records of their earlier investigations, they identified a large number of men who would have been between the ages of fourteen and forty at the time of the first murder and who lacked a credible alibi for the time frame of either crime. They "invited" all those men, more than four thousand of them, to provide a sample of their blood for testing. To save the time and expense of performing DNA testing on all four thousand men, they would first isolate those with the PGM1 + Group A serological type, statistically no more than 10 percent of the sample.

But Jeffreys did not have a chance to do the work. One of the group, a man named Colin Pitchfork, had asked a co-worker to take the blood test for him. Pitchfork's deception was discovered when his co-worker told a friend, and the friend in turn told the police. When the police confronted Pitchfork at his home, he confessed to the two murders. His DNA matched that found in the victims. He was convicted of the two rapes and murders and sentenced to life in prison.

After the first few successes, Jeffreys's lab was inundated with requests for DNA typing. Soon the flood of requests was much greater than the lab could handle, and in almost no time several commercial laboratories stepped into the gap by performing DNA typing for criminal and paternity cases. By 1989 the FBI had become convinced of the value of the new technique and opened its own DNA laboratory.

The commercial laboratories developed tests that were simpler to standardize and interpret, using four or five probes (irradiated fragments), each of which bound to only one part of the DNA molecule. For analyzing DNA in sexual assault cases, where the material collected is often a mixture of sperm and vaginal cells, they developed ways of separating the male and female DNA and analyzing them separately.

Even with only four probes, the commercial labs claimed that the DNA patterns produced were so individual that they were shared by only one in millions of people. In a criminal case this could be strong, even overwhelming, evidence of guilt.

Because DNA typing used such complex technology and claimed to produce incontrovertible identifications, it was carefully scrutinized by the scientific community. Unlike most forensic techniques, DNA identification emerged directly from methods used by scientists in medicine, biochemistry, and genetics. Scientists from these fields were able to point out the shortcomings of the testing methods used by the commercial laboratories as well as the fallacies in the astounding statistics they reported. In spite of its reputation for infallibility, there are indeed weaknesses in DNA testing.

One double strand of human DNA contains about three billion base pairs. Yet forensic DNA typing counts DNA fragments from only a few regions of this enormous molecule. In that sense, DNA typing is unlike fingerprint identification. Where fingerprint examiners can compare the entire print of a suspect to a large fraction of an unknown print, DNA technicians compare only a few scattered pieces of DNA. So far, the sequencing of the entire DNA molecule of every sample is too expensive for forensic use. So all the estimates of the frequency of DNA in the population are actually calculations of the frequency of occurrence of a very limited number of bands.

While it is probably true that no two people on earth share an entire DNA sequence, the likelihood of there being no two people with the same small DNA fragments is not known.

Another question was whether the testing methods originally used on pristine laboratory samples would work on the kinds of samples found in crime scenes—samples in which the DNA might be decomposed or contaminated with DNA from the place where they were found, from the person doing the collecting, or even from the lab itself. Could a lab get a reliable result from a bloodstain found on a street, a dirty carpet, or a pair of jeans? Contamination from material around the DNA sample can prevent the restriction enzymes from breaking up the DNA completely and render the test result unreliable.

Contamination can produce extra bands on the x-ray film. So can leaving the x-ray film on the sample for a longer-than-recommended time. And so can the presence of more than one person's DNA in the sample. How does a lab decide which bands belong to the DNA being tested and which to something else? If there are bands on a DNA profile that don't fit the suspect's DNA, does this mean that the extra bands came from an extraneous source or from another suspect? When DNA decomposes, it breaks up into ever smaller fragments. So, when you break up an old piece of DNA using restriction enzymes, how can you tell whether the resulting pattern is from DNA fragments produced by those enzymes or is the result of decomposition? If a DNA sample is old or small in size, or if a person inherits the same repeating gene sequence from both parents, bands may be missing from the profile. What then should the lab conclude? Can it still be called a match?

No two tests, even under the best of circumstances, produce totally identical results. DNA doesn't always move the same distance up the gel in response to the electric current. For example, a sample with a lot of DNA will move more slowly than one with less. Can you be sure that two DNA samples came from the same person just because the bands moved the same distance on the gel? If they didn't move the same distance, how far apart can two bands be and still be said to match?

Police and laboratories can and do make mistakes. Samples are mixed up or are contaminated with DNA from other suspects, either during collection or in the lab. Sometimes these mistakes are discovered; at other times, particularly when the entire sample is used up in one test, they are not.

If a crime scene sample contains a mixture of DNA from the blood of two or three people, testing can determine whether a sample contains male or female DNA or a mixture of both. And in samples from sexual assaults, sperm can be physically separated from vaginal cells before testing. But in a mixture of DNA where the individual contributions cannot be separated, it is often impossible to distinguish which bands in a sample came from which contributor. If some of the bands in a mixed sample match the suspect in a case but others do not, can you really say that the matching bands all came from the suspect rather than from some other source?

Even when a DNA pattern is clear, calculating the frequency of its occurrence in the world population is not a straightforward operation. Some forensic labs say that if you determine how frequently each band occurs in the general population, you can simply multiply these occurrences to arrive at a probability that one person will have those bands in those particular places. For example, when throwing dice you have a one in six chance of throwing a six on one of the dice, and the same odds on the other. When you multiply the probabilities ($6 \times 6 = 36$), you find that you have a 1 in 36 chance of throwing two 6s on one throw.

According to population geneticists, the frequency of genetically determined traits such as hair, skin, eye color, blood type, and certain diseases varies significantly between different racial and ethnic groups, meaning that the DNA that codes for these variations is also group-dependent. Genes are not always inherited independently from one another. People with light hair and light skin are much more likely to have blue eyes than people with dark hair and dark skin. It would not be accurate to say, therefore, that if one in ten people has blond hair and one in ten has blue eyes, you can multiply the frequency of occurrence of those two genes and get an accurate figure for the frequency of blond-haired, blue-eyed people in the population. And people within an ethnic group tend to marry people in the same group, meaning that it is likely that the genes of a person's parents would have many similarities. Simply multiplying the bands based on these factors would therefore give a fallacious result.

Much research needed to be done before it could be confidently stated that the likelihood of two people sharing the same

DNA profile on a commercial laboratory test was as small as the laboratories claimed it was. DNA testing was litigated in the courts for years, both over the accuracy of the tests and over the statistics used to interpret them. Continuing research was conducted to determine the effects of contamination on the tests as well as on the distribution within different racial and ethnic groups of the DNA fragments used in forensic testing. The National Research Council, a prestigious organization of scientists, published reports in 1992 and 1996 on forensic DNA typing. The reports addressed the questions raised about the accuracy of DNA and whether these had been adequately answered by research. Eventually courts around the world decided that enough was known about how the tests worked and how frequently the patterns occurred to allow DNA test results to be routinely considered as evidence.

Meanwhile the tests themselves were changing. Simpler and more powerful techniques were developed. The most authoritative test, called the RFLP test (Restriction Fragment Length Polymorphism), had limitations that made it less than ideal for forensic purposes. It required a large sample of DNA and took six to eight weeks to complete. Other tests were available that worked more rapidly and on smaller samples, but they did not yield the same stunning statistical outcomes. The other tests might produce a match probability in the range of one in a thousand or two, but not one in a million or even a hundred million as in RFLP testing.

To create tests that would give better statistical results on smaller samples, scientists began working with a set of smaller repeating DNA fragments called short tandem repeats, or STRs. These smaller fragments were amplified using a then-new technique called polymerase chain reaction, or PCR.

PCR uses enzymes to break a length of DNA at each end and split the AT and CG bonds in that length to make two complementary strands. When this is gently heated in a bath of A, T, C, and G molecules, the loose molecules bind to the split strands to create two whole pieces of DNA. The enzyme splits them apart again, and more molecules bind to them to create more strands of DNA, and so on until the quantity of DNA has grown to hundreds or thousands of times the original amount. PCR allows labs to obtain profiles from very small samples. Some analysts claim results can be obtained from a sample with as few as nine DNA molecules.

With PCR available to amplify collected DNA, testable amounts of DNA can be taken from cigarette butts, licked envelopes, clothing, soda-can rims, and even fingerprints. Testing methods have been developed that allow for the testing of up to fourteen areas on the DNA molecule. Instead of band patterns, these new methods yield a computer printout that shows a pattern of color-coded peaks and valleys. Each peak represents an STR sequence on the DNA molecule, one each from the mother and father of the person from whom the sample came. Profiles in which the likelihood of a match is one in tens of billions are common. And the test can be completed in two days. STR has become the name commonly used for the test method itself.

Current STR testing methods are highly automated. First a test sample is amplified using PCR. As part of the amplification process, primer sequences containing a fluorescent dye are added to the mix and bind to the DNA fragments. To help distinguish the primers from one another, four different colors of dye are used. The amplified DNA sample is then loaded into a specialized machine that completes the test. In a process called capillary electrophoresis, the machine uses an electric current to "pull" the DNA sample through a polymer solution in an extremely narrow "capillary" tube. The smaller DNA fragments move more quickly through the tube, the larger ones more slowly.

One section of the capillary tube is constructed of clear plastic, and the machine aims a laser light at that place. As each dyed primer moving through the tube reaches the clear section, it fluoresces under the light. Sensors measure the brightness and color of the fluorescence and transfer the information to a computer, where specialized software translates the light patterns into a printout of peaks of various colors.

Whatever concerns the scientific community had about the science behind STR testing were quickly brushed aside, and today STR profiling is universally accepted as courtroom evidence. Nevertheless some problems with the tests remain. The amplification of such small amounts of DNA can also amplify contaminants, creating extra peaks in the printout; mutations may result in patterns that contain extra peaks; peaks can be produced by the test process itself and sometimes for no known reason. Not everyone agrees on how low a peak must be before it should not be classified a peak; and small quantities of DNA as well as old DNA may yield incomplete profiles with very small peaks or none at all. DNA lab scientists say they can tell most false peaks from real ones by their shapes. But the difference is not always obvious.

When a profile is not an obvious match and requires interpretation to distinguish real peaks from false ones, it is all too easy for the analyst to see what he expects or wants to see in the profile. On occasion a DNA lab analyst has reviewed a printout of DNA evidence containing some peaks consistent with a suspect's DNA, some peaks in areas inconsistent with the suspect, and no peaks at some points where the suspect's DNA would have shown them. The analyst has called it a match, rationalizing that the inconsistent peaks were caused by contamination or accidents, and the missing peaks dropped out of the sample.

Because of the extremely tiny quantities of DNA involved, contamination is an even larger problem with STR testing than with the older tests. Several cases have been reported in which a DNA result has identified the wrong person because of contamination in the laboratory. Often the wrongly identified "suspect" is one of the lab workers. In one instance in the Midwest, the testing of semen from a decades-old rape and murder case resulted in a match with a known sex offender in the area, one whose DNA was being processed in the same government lab for inclusion in an offender database. But for the fact that the identified suspect was only four years old and living in another city at the time the crime was committed, the error might never have been discovered. An adult suspect might have ended up serving prison time for an offense he did not commit.

In a widely reported case in Australia, DNA found on the body of a murdered toddler was matched to a mentally disabled woman living hundreds of miles away. Her family insisted that she had never left her hometown. It turned out that the woman had been sexually assaulted and that the samples from her case had been analyzed on the same day as the samples from the case of the murdered child.

Mixtures are still a problem. Even with the new techniques, bands from multiple sources of DNA cannot generally be distinguished from one another in a printout. New test methods are being developed, however, that will supposedly make it possible to tell which of several people left the DNA in a mix of profiles.

DNA typing of animals and even plants has also been used to solve crimes. Animal DNA has been used to convict poachers of endangered animals, birds, and fish as well as smugglers of prohibited animal products. DNA typing of dog and cat hair has been used to link animal hair found at a crime scene with the suspect's pets. In at least one case, DNA typing of plant material found in the defendant's truck was used to show that the truck had been in the area of the crime.

The success of RFLP and STR testing has further encouraged forensic scientists to look at other forms of DNA typing being used in biology and medicine.

Mitochondria are often called the energy factories of cells. There are as many as two thousand microbe-sized mitochondria in the cells of every animal, plant, and fungus. They reside outside the nucleus of the cell, have their own separate membranes, and convert sugar and oxygen into a compound called adenosine triphosphate (ATP). ATP provides the energy that allows the cells and the organisms they comprise to do what they do. Mitochondria are passed from mother to child, and, except for random mutations, remain unchanged through the maternal line for many generations.

The DNA found in mitochondria is unique to the mitochondria and bears no relation to the cell's own DNA. The mitochondrial DNA genome was first sequenced in 1981.

Our bodies contain a great deal more mitochondrial DNA than nuclear DNA. Mitochondrial DNA testing has become popular with archaeologists and anthropologists because testable amounts of mitochondrial DNA may continue to exist in ancient bones and teeth long after the nuclear DNA has disintegrated. Scientists have been able to obtain sequences of mitochondrial DNA from samples that are thousands of years old. Mitochondrial DNA testing can also be done on a material such as rootless hair, which contains no nuclear DNA.

In forensics, mitochondrial DNA typing may help make an identification when a specimen is too old or degraded to permit even STR typing.

Mitochondrial DNA typing is more difficult to perform than STR typing, and the results are less discriminating. Because we receive our mitochondrial DNA only from our mothers, any person's mitochondrial DNA is the same as that of his or her mother and of all his or her maternal relations. So many people can have the same mitochondrial DNA, in fact, that it is not as conclusive as STR typing result for identifying the source of a sample.

Mitochondrial DNA testing has been used in the aftermath of wars and oppressive political regimes to identify bodies in mass graves. In 1993 it was used to help solve the mystery of the disappearance of the Russian royal family during the Bolshevik Revolution. The Romanovs—Tsar Nicholas II, his wife Alexandra, and their five children—were imprisoned and killed by the Bolsheviks in 1918. But the location of their remains was not revealed until 1991, after the fall of the Soviet Union. Their bodies were found in an unmarked grave in a forest near Yekaterinburg.

Sulfuric acid had been poured on the bodies after the murders, and the bones had been moved from their original site and reburied. Besides the incomplete skeletons, little was left with which to identify the bodies. To confirm that the bones found at the site were in fact those of the Romanovs, researchers used both STR and mitochondrial DNA testing. STR testing of tiny amounts of nuclear DNA recovered from the bones revealed that they had come from a family—a father, mother, and three daughters—as well as from five unrelated persons. According to records, the additional remains were probably those of four servants and the family doctor, killed along with the tsar and his family.

Mitochondrial DNA was recovered from the bones, sequenced, and compared to that of known maternal relatives of Nicholas and Alexandra, a group of European aristocrats that included England's Prince Philip. The mitochondrial DNA in the bones believed to be the tsar's were mostly a match with Nicholas's relatives, but they did show a rare mutation that was not present in any of the living relations. To confirm the match, researchers recovered mitochondrial DNA from the body of Nicholas's brother; testing showed the same mutation.

More bone fragments found near the burial site in 2007 were also tested and appear to be those of the remaining two children.

Mitochondrial DNA testing was also done on a tissue sample from Anna Anderson, a woman who, until her death in 1984, claimed to be the youngest Romanov daughter, Anastasia. The tests confirmed that she was not a member of the Romanov family.

Mitochondrial DNA testing was also used to identify the Vietnam War–era soldier whose remains were placed in the tomb of the unknown soldiers in Washington, D.C. The remains were believed to be those of air force lieutenant Michael J. Blassie, but this identification was not confirmed until 1998 when mitochondrial DNA was extracted from his bones, tested, and compared with that of Lieutenant Blassie's mother.

A recent addition to forensic DNA typing, Y-STR testing, was made possible by the sequencing of the Y chromosome of the human genome. The X and Y chromosomes determine the sex of a human being: women's DNA has two X chromosomes while men's has an X and a Y. Since only men have Y chromosomes, DNA typing using short tandem repeats unique to the Y chromosome (Y-STR) can detect and identify male DNA in a mixture. In samples in which the DNA of more than one male is present, it can also help distinguish one profile from another.

Standard STR typing can also detect the presence of male DNA using a sequence on the amelogenin gene that has been shown to be different in men and women. But detecting male DNA with this marker can be difficult when very little of it is present in a sample.

The usefulness of Y-STR typing is limited because, like mitochondrial DNA, it is transmitted by only one parent. Since women do not have a Y chromosome, a man's Y chromosome must come from his father. And, unless it mutates, this DNA will be identical to that of his father, grandfather, paternal uncles, sons, and so forth—all his paternal relatives.

Y-STR typing has been used in genealogy to determine relatedness. Perhaps the most famous example of Y-STR typing was its use in determining if Thomas Jefferson fathered the children of Sally Hemings, one of his slaves.

Stories have circulated from as long ago as Jefferson's own lifetime that Sally Hemings was Jefferson's mistress and that some or all of her six or seven children were his. In the 1990s researchers interested in seeing whether the story could be confirmed by DNA typing sought out relatives who could be tested. They were able to find living descendants, in an unbroken male line, of two of Hemings's sons. On the other hand, there are no living descendants in the male line of either Thomas Jefferson or his only brother. But researchers found some male descendants of Jefferson's paternal uncle.

Y-STR testing on the living male descendants of both Jefferson and Hemings showed that one of Sally Hemings's great-great-great grandsons, the descendant of her youngest son, had a Y-STR profile identical to that of Jefferson's uncle—and, by implication, identical to that of Jefferson himself. The descendants of another of Hemings's sons had a profile that differed at several points from the Jefferson family profile, showing that he was not a descendant of Jefferson's.

The results in this case demonstrate both the power and the limitations of Y-STR testing. While the Y-STR profile match is conclusive evidence that one of Sally Hemings's great-great-great grandsons is directly related to Thomas Jefferson, it does not prove conclusively that Jefferson was his great-great-great grandfather. The father of Sally Hemings's son might have been Jefferson, but it might also have been one of Jefferson's relatives, a number of whom spent time at Monticello. The Y-STR typing provided evidence that bolsters the stories about Jefferson's relationship with Sally Hemings, but not enough to prove conclusively that they are true.

Once DNA's power to identify people was recognized, it did not take long to decide that a large-scale database of DNA profiles,

like the AFIS database of fingerprints, could be a major resource
for law enforcement in identifying suspects in criminal cases.
Most states now have laws that allow the police to collect saliva and blood samples from arrestees or convicted criminal offenders. These DNA profiles are then uploaded into databanks.
The FBI also keeps a national database of offender profiles. The
standard method for creating database profiles uses thirteen particular STR markers plus the amelogenin marker for the gender
of the subject.

The FBI's DNA database, NDIS, now contains about three
million profiles. Various state databases altogether contain many
more. As the number of stored profiles has increased, so has the
number of cases in which a suspect has been located by comparing DNA found at a crime scene to that of known criminals in
the offender databases. "Cold hits," as these matches are called,
have resulted in convictions in many previously unsolved cases,
some of them many years old. Because of the success of DNA
databases in identifying perpetrators of crimes, some authorities have proposed expanding them to include the profiles of all
Americans.

More recently, DNA databases have been used to locate suspects through their close relatives. In a case in Great Britain, a
killer whose DNA was not in the national database was located
when blood he left on the murder weapon turned out to be a
close, but not quite complete, match to another profile in the
database. Police turned the focus of their investigation to that
man and eventually found that the killer was his brother.

Privacy advocates have raised alarms about the use of DNA
databanks to harass innocent people who happen to be relatives
of a criminal suspect. And scientists have pointed out that in the
huge offender databases now in existence, even people who are
not related may have profiles that share many similarities. So the
likelihood that profile-searching will actually bring police closer
to a perpetrator may not be great enough to justify the expense

of investigating close matches or the risk of disrupting the lives of the innocent.

As research scientists develop faster, better, and more sensitive methods of detecting, sequencing, and comparing DNA, forensic laboratories continue to look for ways to adapt those methods in order to improve the quality of their own testing. Around the world forensic DNA test methods are constantly being sought that would allow forensic scientists to discriminate among profiles in a mixture and obtain test results from ever smaller amounts of DNA.

Among the forensic sciences, DNA typing is unique in its close relationship to the biological sciences and in the continuing interest of scientists in the forensic uses of DNA typing techniques. The courts have been willing to listen when the scientific community weighs in on the validity of a particular use of DNA typing. This interest has made DNA typing one of the most reliable of the forensic sciences.

On December 23, 1972, Diana Sue Sylvester, a twenty-two-year-old nurse, was raped and murdered in her San Francisco apartment. The police had one suspect, based on a neighbor's identification, but the identification was shaky, and no other evidence was ever developed against the man.

On February 21, 2008, thirty-five years later, a wheelchair-bound seventy-one-year-old handyman named John Puckett was convicted of her murder (the statute of limitations had run out on the rape charge). He had been convicted of rape previously, but there was no known connection between him and Diana Sue Sylvester. The only evidence against him was a DNA "cold hit" on the California CODIS system.

CODIS, the Combined DNA Index System, was set up by the FBI in 1994. Originally intended to keep DNA records of all sex offenders, the index has been broadened to include the DNA profiles of all persons arrested on federal charges, the profiles of missing persons, and DNA samples gathered at crime scenes but as yet unidentified. Most states now have CODIS systems of their own.

During Puckett's trial the prosecution held that although the DNA "hit" was the only evidence against the defendant, it was enough to convict him. They argued that the odds of a coincidental match between Puckett's DNA and the sample found at the crime scene were one in 1.1 million. Based on this argument, Puckett was convicted and sentenced to seven years to life in prison.

The judge, however, had refused to allow the jurors in the case to hear that there was considerable controversy over that one in 1.1 million figure, and that some statisticians believed the figure should be closer to one in three.

There are two ways of looking at the problem presented by Puckett's cold hit. Statistically, the results can be vastly different. The odds of finding a particular match in a database are drastically different from the odds of finding any old match. The "birthday paradox" explains what I mean.

What are the odds of anyone in the room having the same birthday as yours? The size of the group doesn't matter. It could be a small gathering or could include everyone at Grand Central Station at 8:30 in the morning. The odds are in fact one in 365. And the odds of having a better-than-even chance of finding a match are one in 183—half of 365 for a 50 percent chance.

But now let's look at the problem differently and make it nonexclusive. What are the chances that any two people will have the same birthday? How many people must be in a room before the odds are 50-50 that if everyone calls out their birth date, two of them will match? The math is a bit complex, involving combinational statistics, but the number works out to 23. That means that if you take random groups of 23 people and check their birthdays, half the time you will get a match. Just changing it from a specific match—someone else having *your* birthday—to a nonspecific match—any two people having the same birthday—changes the number of people in the group from 183 to 23.

And this is what searching for cold hits in CODIS does—it checks everyone in the system against everyone else in the system. And if you run the same math on CODIS as you do on the birthday paradox, you will find that the odds of a cold hit are one in three. Yes, but what are the odds that the person you find will be a sex offender? The odds of that are 100 percent since that is the nature of the database you are running the test on.

The acceptance of DNA evidence in the courtroom and the belief in its accuracy is now so complete that complacency has begun to set in. Rather than giving any statistics at all, the criminalist on the stand need only say, "The DNA evidence says it's him, so it's him." He may not be telling the court and the jury about technical difficulties—the unexplained allele dropout at position five, or the mysterious peak somewhere else on the chart. It may not be the actual DNA evidence that the court hears but rather the criminalist's interpretation of the DNA evidence. The jury needs to know this.

13 : Bones and Bugs

FORENSIC ANTHROPOLOGY is the study of the human skeleton and of the facts it can reveal about the age, sex, health, physical condition, cause of death, and time of death of the person once intimately connected to it. With luck there will be enough information in the bones to identify the person and, in cases of homicide, to suggest a possible perpetrator.

The man considered to be the father of American forensic anthropology is Thomas Dwight (1843–1911), whom the *New York Times* once called "America's foremost anatomist." Dwight was the Parkman Professor of Anatomy at Harvard from 1883 until his death. Perhaps the source of Dwight's interest in the forensic aspect of his profession was the 1849 murder of the same Parkman who had endowed the chair of anatomy he himself occupied. At the time of Parkman's death, the chair he had endowed was held by Oliver Wendell Holmes, father of the jurist Oliver Wendell Holmes who would become one of the great Supreme Court justices of the twentieth century.

In 1850 John White Webster, M.A., M.D., and a professor of anatomy, had the distinction of becoming the first Harvard professor ever to be hanged for murder. The victim, Dr. George Parkman, has been variously described as a wealthy socialite and philanthropist, or as a rich, vain, bad-tempered, money-grubbing skinflint. It depended, I suppose, upon whether or not you owed him money. Dr. Oliver Wendell Holmes described him as the perfect Yankee: "He abstained while others indulged, he walked while others rode, he worked while others slept." Professor Webster did not have as high an opinion of Parkman. Perhaps this is because Professor Webster owed him money. Webster had put up a valuable collection of gems as collateral for a loan from Parkman. Then Parkman discovered that Webster had put up the same collection as collateral for a loan from someone else. Parkman was miffed.

Parkman went to see Webster in his laboratory in the basement of the medical building at Harvard on the afternoon of Friday, November 23, 1849, to demand his money back. "I haven't got it," said Webster.

Parkman threatened to have Webster fired. And because Parkman had recently donated the land for the new Harvard Medical School, there was a good chance that he could do as he threatened. In a fit of strong emotion, Webster picked up a hunk of wood and struck Parkman in the head, killing him. He dragged the body to a large sink and, in a much calmer frame of mind, proceeded to dissect it. He put the choice bits in collecting jars and lined them up on a shelf. Fragments of bone and such he burned in his assaying furnace; the larger body parts he tossed down an indoor privy in a corner of his lab.

Parkman's disappearance did not go unnoticed. It was assumed that he had been abducted, and posters were hung all around Boston. The university offered a $3,000 reward for information as to his whereabouts or the identity of his abductors. Webster went to see Parkman's brother and volunteered that he might have been the last person to see Parkman alive. He showed the brother a receipt for $483 allegedly signed by Parkman and acknowledging repayment of Webster's debt.

The following Thursday, Thanksgiving Day, Webster made the mistake of giving the college janitor a turkey. The janitor,

Ephraim Littlefield, thought this odd behavior for Webster, who, as far as Littlefield knew, had never given anything to anyone. He had overheard the argument between Webster and Parkman but had left the building before its fatal conclusion. Littlefield decided to investigate the reason for Webster's sudden munificence and sneaked into the basement lab the next day. The wall of the assay oven felt warm to the touch, but the oven was locked. Littlefield, with his wife standing guard, got a hammer and chisel and broke through the brick wall. In the oven he found bones—a human pelvis and several pieces of leg. He called the college authorities, who in turn summoned the police.

Webster was arrested at his home in Cambridge. "That villain!" he cried, when told of Littlefield's chiseling. "I am a ruined man!" He attempted to commit suicide in his jail cell by taking strychnine, but he survived.

By the time of the trial Webster had recovered his composure and devised an alibi. The bones, he claimed, were the remains of a cadaver, a medical school specimen he was disposing of.

Then, for the first time in judicial history, forensic anthropology and forensic dentistry were called upon to make the case for the prosecution. Oliver Wendell Holmes, the Parkman Professor and dean of the Medical College, testified that all the various body parts found in Webster's basement were "consistent" with Parkman's anatomy. Dr. Nathan Keep, a dental surgeon, testified that the teeth found in the furnace were the lower left portion of a set of teeth he had made for Parkman three years earlier. As it happened, he had kept the mold. When the teeth were compared to the mold, they were a perfect match.

The jury took three hours to find John White Webster guilty of the murder of George Parkman. A few days before his execution six months later, Webster wrote a confession, saying in part:

. . . I was excited to the highest degree of passion; and while he was speaking and gesticulating in the most violent and

menacing manner, I seized whatever thing was handiest—it
was a stick of wood—and dealt him an instantaneous blow
. . . on the side of his head. . . . He fell instantly upon the
pavement. There was no second blow. . . . Blood flowed
from his mouth and I got a sponge and wiped it away. I got
some ammonia and applied it to his nose; but without effect.

The first thing I did . . . was to drag the body into the
private room adjoining. There I took off his clothes, and
began putting them into the fire which was burning in
the upper laboratory. They were all consumed there that
afternoon—with papers, pocketbook, or whatever else they
may have contained.

My next move was to get the body into the sink which
stands in the small private room. . . . There it was entirely
dismembered . . . as a work of terrible and desperate
necessity. The only instrument used was the knife found by
the officers in the tea chest, and which I kept for cutting
corks.

The case is notable for two reasons: it was the first time an
American court had considered scientific testimony; it was also
the first time that a Harvard professor was hanged.

Adolph Louis Luetgert (1845–1899) came to America from Germany in the 1860s and settled in Chicago. After a series of menial
jobs, in 1879 he started his own sausage-manufacturing company. A year earlier his first wife had died, and he had then married Louisa Bricknese, a petite, charming girl ten years his junior.
To sanctify the wedding, Luetgert gave his new bride a heavy
gold wedding band engraved with the initials "LL."

According to Luetgert, on May 1, 1897, Louisa went to visit
her sister. She never returned. As it turned out, she had never
arrived. Luetgert went to the police to report his wife missing.

But when she remained away for several more days her brother began to suspect that she was more than merely missing, and that Luetgert might know something about it. The couple had been fighting bitterly for the past few years, so much so that Luetgert had set up a bed in his office at the factory. Rumor had it that Luetgert had been ogling a rich widow.

It occurred to the police that a sausage factory might be the ideal place to dispose of a human body. Indeed, when they searched the building they found a huge vat filled with a caustic solution. They drained the vat and found nothing but four tiny bone fragments and two rings. One of the rings was a thick gold band with the initials "LL" engraved on it.

His wife had visited the factory many times, Luetgert told them. She may have dropped it in the vat by accident.

And the bones?

Sheep bones, pig bones—after all, this was a sausage factory.

Luetgert was tried for murder in August 1897. His defense, combined with reports of sightings of Louisa Luetgert from all around the country, was enough to create a hung jury.

But in January 1898 Luetgert was retried. This time the prosecution called in George A. Dorsey, one of the few expert forensic anthropologists of the time. Dorsey examined the four bone fragments, each smaller than a dime, and declared them to be human. One was the tip of a rib, one a bit of phalanx (toe bone), one a sesamoid bone from the foot, and one the end of a metacarpal (one of the bones that connect the fingers to the wrist).

This time Adolph Luetgert was convicted of the murder of his wife and sentenced to prison for life. Later that year George Dorsey published an important paper, "The Skeleton in Medico-Legal Anatomy," based on what he had learned while performing research for the trial.

Not everyone agreed with Dorsey's conclusions. In an unsigned article in the *Medical News* for March 1899 ("Medical

Matters in Chicago"), the correspondent attacked Dorsey's description of each bone, concluding that "Just as well might one swear to the identity of a coat, when every intrinsic vestige of the garment has disappeared except a button-hole." But even if the correspondent was right, there was still that damn ring.

On September 7, 1935, Dr. Buck Ruxton, a Parsee who had been born in India in 1899 and educated at Bombay University, accused his "wife" of having an affair with the town clerk. She was Isabella Van Ess, whom he had never gone through the formality of actually marrying. The town was Lancaster. Ruxton had moved to England and gotten a second medical degree from the University of London.

One week later Isabella disappeared along with her maid, Mary Rogerson. The next day, Sunday, September 15, Ruxton told his charwoman, Agnes Oxley, not to bother coming on Monday as his wife was away. On Monday morning he took his children to a friend's house for the day. On the way home he stopped at the home of Mary Rogerson's parents and told them that Mary was pregnant. His wife had taken her to Scotland, he said, in order to "get this trouble over." Everyone who came to the door of his house at 2 Dalton Square seeking medical attention that day was sent away. He was, he said, putting in new carpets. That evening he rented a car, and on Tuesday he was away all day.

Two weeks later, on Sunday, September 29, the mutilated and decomposed bodies of two women were found under a bridge near Moffat, Scotland, 107 miles from Lancaster. The remains were scattered about a ravine known locally as the Devil's Beef Tub.

The police added up the facts, and on October 13 they arrested Ruxton for the murder of Mary Rogerson. Three weeks later they added the murder of Isabella to the charge. Ruxton called the charges "absolute bunkum, with a capital B."

Two women were missing under suspicious circumstances, and two bodies had been found. The charwoman told of new stains on the floor and a foul odor in the house when she had come to clean for Ruxton on that Tuesday. A Mrs. Hampshire, who had been hired to clean the staircase, said that the water ran red when she scrubbed the carpet. All very suggestive, but not proof of anything.

The police somehow had to connect Ruxton with two skulls, two torsos, and an assortment of limbs and soft tissue. Various teeth had been pulled from the skulls to make dental records useless for identification. Isabella's fingertips had been cut off, so that fingerprints could not be used to identify her. Mary Rogerson had never been fingerprinted and so her hands were intact. But the eyes were missing from her skull, so that her pronounced squint would not be in evidence. The tissue had been removed from one pair of legs—Isabella's legs were noticeably chubby. Isabella had had a bunion on one foot. Indeed, from one of the severed feet where a bunion might have been, a piece of flesh had been removed. Again, this implicated Ruxton but was not proof of anything.

Two forensic experts, Dr. John Glaister of Glasgow and Dr. James Couper Brash of Edinburgh, were called in to see if they could somehow identify the bodies. They managed to pick up a few small details that Ruxton had missed. Mary Rogerson suffered from recurring tonsillitis, and the doctors found microscopic signs of the disease in the tonsils of one corpse. The missing bunion was connected to a deformity in the bones of the severed foot. Missing teeth that had been pulled before death were matched with dental records.

As a clincher, the forensic experts used pictures of the missing women and superimposed them on photographs of the skulls taken at exactly the same angle. The skulls fit inside the photographs exactly. It was a convincing demonstration, and the jury

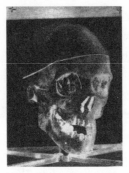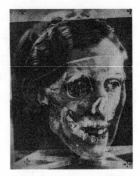

In the Ruxton murder case, forensic experts superimposed photos of Isabella Van Ess and the skull of a decomposed body to prove her identity as one of the victims.

found Dr. Buck Ruxton guilty. On May 21, 1936, he was hanged at Strangeways Jail in Manchester. In 1937 Drs. Glaister and Brash published a book about the case, *Medico-Legal Aspects of the Ruxton Case*, which detailed both the police work and the scientific investigations. Reviewers were impressed by the extreme care taken by the scientific examiners, and the book helped set standards for future investigations.

The Anthropology Research Facility at the University of Tennessee in Knoxville was founded by Dr. William M. Bass in 1972 when he realized just how little was known about how human bodies decay after death. At the time, Bass was the official state forensic anthropologist, and he did not like being asked questions about human remains that he could not answer. The police often had such questions. Bass got permission to use a two-and-a-half-acre plot of land owned by the University of Tennessee; then he gathered unidentified and unclaimed bodies from the local medical examiners' offices and distributed them about the property. Some were left out in the open, exposed to the elements and insects, some were buried in shallow graves, some were left in car trunks, and some were partially or completely submerged in water. Records were kept that detailed the progress of each body's decomposition.

Soon after its founding, the facility acquired the nickname "The Body Farm." The name stuck. The Body Farm is still in operation and receives about 120 donated bodies every year. In addition to its scientific fact-gathering, police departments use the facility to train personnel in scene-of-the-crime exercises.

There are now body farms at Western Carolina University and Texas State University, and several other institutions are considering opening similar facilities. The study of the defunct human body, though a bit macabre, provides useful information and helps make forensic anthropology a more exact science.

Who saw him die?
    "I," said the fly.
"With my little eye,
    I saw him die."
—"Cock Robin," English folk song

The earliest-known case of insects helping to solve a crime is the story told by Sung Tz'u, recounted in Chapter One, of blowflies gathering on a peasant's sickle and thereby indicating that he was a murderer. Except for this one thirteenth-century case, forensic entomology lay dormant until 1855 when Dr. Louis François Etienne Bergeret (1814–1893) performed an autopsy on the mummified remains of a baby. The child's body had been found behind the mantelpiece of a house near Paris as the house was being remodeled. Bergeret used his knowledge of the life cycle of the insects whose remains he found in the body to conclude that the death had occurred more than five years earlier. He settled on 1848 as the year of the infant's death, and the young woman who had lived in the house that year was found, arrested, and tried.

She was not convicted because it had been impossible to show that the child had not died a natural death. And perhaps it had.

In several cases in the late nineteenth century, the parents of dead babies were spared criminal charges when injuries to the faces and bodies of the children were shown to have been caused postmortem by swarms of roaches.

Entomological knowledge has only recently become systemized enough to be useful in solving crimes. In its life cycle an insect goes through many stages, and each stage may leave behind a telltale trace. Insects can be amazingly precise about where they deposit their eggs, the temperature of their chosen spot, and the time of year or even the day when they lay their eggs. Each species of insect has its own specific rules of conduct about such things.

Two major insect groups, flies and beetles, most commonly colonize corpses, and forensically valuable information can be had from determining the order of their arrival and departure from their hosts.

Each stage in the decomposition of a cadaver is attractive to a different insect. Blowflies (*Calliphoridae*) and houseflies (*Muscidae*) are usually the first to arrive at a fresh cadaver, though occasionally flesh flies (*Sarcophagidae*) will also make an appearance. One or another of these will often show up within minutes of death. The female blowfly will lay her eggs within two days, preferably in a wound, though any body orifice will do. Later come a variety of insects attracted by the changing degrees of decomposition of the corpse or by the insects already there—rove beetles (*Staphylinidae*) may show up to feed on maggots, for instance. The moldering corpse may attract picture-winged flies (*Oititidae*), soldier flies (*Stratiomyidae*), or a variety of others. The last to appear are usually the hide beetles (*Dermestidae*) and hister beetles (*Histeridae*). The order in which insects appear is a function of the geographical location of the body and the time of year.

For the first few weeks after death, information as to how long the corpse has been dead and has lain where it was found can be determined from the age and developmental stages of the blowfly maggots in the body. An examination of the pupal cases—the hard shells that are left behind when the insect changes from its final larval stage to its adult form—can provide the investigator with an estimated period of death.

Traces of insects that are not native to the area in which the body was found are an indication that the body has been moved. Insect populations change even over very short distances. The interior of a house or barn will provide a different insect population than the exterior.

Another gift the insect may give the investigator is information about the poisons or drugs to be found in the tissues of the corpse. Any chemicals found in the insect surely came from the corpse it was feeding upon.

Forensic entomologist Dr. Zakaria Erzinclioglu describes a case to which he was called in West Yorkshire, England, in the 1980s. Anthony Samson Perera, a lecturer in oral biology at the dental school of Leeds University, lived with his wife in a house in Wakefield with two small sons and an adopted thirteen-year-old daughter named Nilanthi. The neighbors, who were fond of the daughter, one day realized that they had not seen her for quite a while. This was odd because it was midsummer, not a time for a thirteen-year-old to be spending the entire day indoors. When they asked the Pereras about her, they were told at first that she was indeed inside the house. But Nilanthi remained invisible, and the neighbors notified the social services department that they were worried about her.

The social services people turned the case over to the police, and Detective Inspector Tom Hodgson soon went to call on the Pereras. "The first thing he noticed," according to Dr. Erzinclioglu, "was that Dr. Perera was an insufferably arrogant man. His

manner towards Hodgson and his colleagues was one of conde-
scension, bordering on the ill-mannered." Perera told Hodgson
that the missing girl was an orphan whom they had adopted in
Sri Lanka, where they came from. They had brought Nilanthi
with them to England so that she could go to school, but the girl
was unhappy there and had gone back to Sri Lanka where she
was now living with Perera's mother, Winifred. Perera couldn't
understand why such a fuss was being made over "a simple jun-
gle girl." And how had Nilanthi been returned to Sri Lanka? Perera had flown with her to Sicily, he said, and had turned her over
to his brother, who presumably had taken her to Sri Lanka when
next he went home.

Inspector Hodgson checked with the airline; they had no record of Nilanthi's ever leaving England. He asked Interpol to
query the Sri Lankan police. They reported that Mrs. Winifred
Perera told them that she had not seen Nilanthi since the girl left
for England three years before.

Perera shrugged it off. "Ask my brother," he said. And where
was his brother? He had overstayed his Italian residence permit
and had had to leave Italy. Where he was now, Perera didn't
know.

But Perera's tremendous conceit and his apparent belief in his
superior intelligence led him down a dangerous path. He brought
some bones into his laboratory, put a few of them into glass jars,
and left the others lying about in enamel dishes. One of his colleagues, Frank Ayton, called Inspector Hodgson. Ayton showed
Hodgson a coffee jar holding a few small bones, and a five-liter
beaker filled with bones floating inside in a green liquid. The two
of them searched the laboratory and found a stainless steel tray
that held still more human bones.

On the basis of these finds, Inspector Hodgson obtained a
warrant to search Perera's house. The searchers found human
remains in three indoor plants and more human bones under

the floorboards in the living room. They found yet more human bones and a quantity of long black hair in the garden.

Perera's unshakable confidence and arrogance did not desert him even in the face of these finds. The human bones were merely biological specimens for his studies, he said. They were under the floorboards to avoid any "misunderstandings" when he had guests. And the bones and rotting flesh in the plant pots were from pork chops used to fertilize the plants. Oh, they were definitely human? Well, they must be from the cadaver he had ordered from Peradeniya University in Sri Lanka.

Despite the glib assurance of his answers, Perera and his wife were placed under arrest. But all this occurred in the days before DNA typing, and the police needed something to tie all those bones together. They bundled everything up and brought it to Dr. Erzinclioglu's laboratory in Cambridge. They wanted to know if the bones from the garden were connected to the bones in the pots and under the floorboards.

Dr. Erzinclioglu held up the plastic bag holding the bones from under the floorboards. It was teeming with "hundreds of tiny mites belonging to species that are predatory upon other small invertebrate animals in the soil." But the area under the floorboards was concrete; there was nothing there for the mites to feed on. So the bones must have been brought in from the garden. And there was more. The bones from Perera's lab held insects that lived in houses, not labs. And some of the larvae were from a species that begins breeding in the spring—around the time that Nilanthi disappeared.

The bones were examined by other experts who concluded that they certainly could be the remains of a thirteen-year-old girl of Asian ancestry. But they could not be more positive than that.

To address the final possibility that Perera was somehow telling the truth—that Nilanthi was alive and well and living with Perera's mother—Hodgson traveled to Sri Lanka. Nilanthi, it turned out, was not an orphan. But her parents had neither seen nor heard from her since the day she left for England with Perera. And Peradeniya University had never sent Perera a cadaver.

As he sat through the trial, Perera lost none of his arrogance. As he saw it, nothing could be proved against him; it was all supposition and circumstantial evidence. But the weight of the circumstantial evidence proved overwhelming. Nilanthi had certainly come to England with Perera. She was certainly no longer there. She was not in Italy or Sri Lanka. Where was she? The bones in Perera's house, garden, and laboratory were human. They could have been those of a thirteen-year-old girl. The insect evidence showed they had been put in the ground at the same time that Nilanthi disappeared.

Perera was found guilty and sentenced to life in prison. His wife, against whom no evidence had been introduced and who seemed genuinely bewildered by the circumstances, received a suspended sentence and was released.

14 : The Eyewitness and Others

It's the wrong time and the wrong place.
Though your face is charming,
it's the wrong face.
It's not his face, but such a charming face
That it's all right with me.
—Cole Porter

There had been an automobile accident.
Before the court one of the witnesses, who
had sworn to tell "the whole truth, and
nothing but the truth," declared that the
entire road was dry and dusty; the other
swore that it had rained and the road was
muddy. The one said that the automobile
was running very slowly; the other, that
he had never seen an automobile rushing
more rapidly. The first swore that there
were only two or three people on the village
road; the other, that a large number of
men, women, and children were passing
by. Both witnesses were highly respectable
gentlemen, neither of whom had the
slightest interest in changing the facts
as he remembered them.
—Hugo Münsterberg,
*On the Witness Stand: Essays on
Psychology and Crime,* 1906

WHEN A VICTIM or a witness to a crime points a finger at a suspect and cries, "He's the one! I'm sure!", what are the odds that the accuser is right?

Eyewitness testimony is notoriously unreliable. Prosecutors know it, and so do defense attorneys. Then why is there so much of it? Because juries do not know, and the sight of the victim sitting in the witness box and pointing that quavering finger at the defendant is a powerful psychological boost to the prosecution's case. Even if you know that eyewitness testimony is often unreliable, it's hard to believe that this slender girl who has been horribly molested, or this elderly, frail man on crutches with his head still swathed in bandages, could be lying.

And, of course, they're not lying. Not consciously, anyway. Consider the case of "Lord Willoughby" in Chapter 3. The ladies who appeared in court against Adolf Beck honestly believed that they had identified the right man.

In his 1908 book *On the Witness Stand*, Dr. Hugo Münsterberg discusses perception and memory as they pertain to eyewitness testimony. In one of his chapters, "The Memory of the Witness," he recounts his own experience in giving testimony in a burglary case. He appeared in court not as an expert witness but as a victim. While he and his family were away at the seashore, his Boston house had been burglarized.

Münsterberg testified that the burglars had entered through a cellar window and moved about upstairs from room to room. He told the court that he had found drops of candle wax on the second floor, indicating that they had been there at night. The burglars had left a large mantel clock wrapped in paper on the dining room table, showing that they had intended to return. They had also taken some articles of clothing, and Münsterberg gave the police a list of the missing items.

Within a few days after he appeared on the stand, Münsterberg realized that his testimony had been wrong in every detail. The burglars had not come in through a window, they had broken the lock on the cellar door. The clock was on the dining room table, but not wrapped in paper—it was covered with a tablecloth. The candle had dripped not on the second floor but in the attic. And the list of missing clothing was incomplete by seven garments.

And this from a psychologist, a criminalist, and a trained observer with an excellent memory.

In his book, Münsterberg goes on to analyze why he made these mistakes: he was in a high state of excitement (he had rushed home when the police told him that his house had been broken into); he had focused too much on a specific observation (the clock, the wax) without really noticing the secondary details (what the clock was actually wrapped in, and on which floor the wax was actually spilled). When it came time to tell the story, his mind had then supplied the supporting details. As for the cellar window, the police had mentioned that burglars usually enter through the cellar window; so Münsterberg assumed that this is what he had seen. It was not until a few days later that he noted that the lock on the cellar door had been forced.

The Survival, Evasion, Resistance, and Escape programs of the United States Army conduct experiments to find out how best to prepare soldiers in the event they become prisoners of war. One of these experiments, devised by Yale psychiatrist C. Andrew Morgan, has uncovered a fascinating statistic. In the experiment, a test subject undergoes a tough interrogation and is put into an isolation cell. Later, another captor bursts into the cell. He waves a photograph and asks, "Did your interrogator give you anything to eat?"

That's it. That's the extent of the misdirection.

As the experiment comes to an end, the subject is asked to pick out his interrogator's face from among nine photographs he is shown. An amazing 85 percent of the subjects choose the wrong face. Instead of the face of their actual interrogator, they pick the one they had seen briefly in the waved photograph.

Eighty-five percent.

Professor Elizabeth Loftus holds appointments at the University of California, Irvine, in the departments of Psychology and Social Behavior and Criminology, and in Law and Society and Cognitive Sciences as well as at the Center for the Neurobiology of Learning and Memory. Loftus has been called upon to consult or testify on questions of memory in hundreds of cases, including the McMartin Preschool molestation case, the Hillside Strangler case, and the Oklahoma City bombing case. She is an expert—*the* expert, in fact—on memory and what can be done to distort it. Her main finding, after three decades of research, is that it does not take much. In a 2002 article in *Skeptical Inquirer*, she wrote, "A memory is, of course, not proof of the event it purports to recall. We all 'remember' things that never actually happened, as ample scientific evidence has demonstrated."

Eyewitness testimony, Loftus finds, not only is unreliable but can be influenced by the way questions regarding the incident are phrased. In one study she showed her subjects films of traffic accidents and then asked questions. If she asked how fast the cars were going when they *smashed into* each other, the speeds reported by her subjects were higher than if she merely asked how fast they were going when they *hit* each other. *Smashed into* also resulted in the subjects' recalling broken glass when in fact there had been none.

The ways of inducing misinformation into memory are myriad. Subjects have been convinced that a stop sign at an intersection was a yield sign, that a blue car was white, and even that Mickey Mouse was Minnie Mouse.

In addition to its innate unreliability, one of the seldom-stated problems with eyewitness testimony is the habit of the police and prosecuting attorneys of cherry-picking their eyewitnesses. If there are three eyewitnesses to a crime and one says the perpetrator had brown hair, one says blond, and the third says he was bald, and the defendant on trial has brown hair, which of the three witnesses do you suppose will testify for the prosecution? In theory the prosecution is supposed to turn over all evidence, both inculpatory and exculpatory, to the defense. But all too often in practice a great deal of information the defense might find useful is discarded along the way. In 1924 J. Collyer Adam, in his introduction to Hans Gross's book *Criminal Investigation*, addressed the difficulty of eyewitness testimony.

> It must be admitted that at the present day the value of the deposition of even a truthful witness is much overrated. The numberless errors in perception derived from the senses, the faults of memory, the far-reaching differences in human beings as regards age, sex, nature, culture, mood of the moment, health, passionate excitement, environment—all these things have so great an effect that we scarcely ever receive two *quite similar* accounts of one thing; and between what people really experience and what they confidently assert, we find only error heaped upon error.
>
> Out of the mouths of two witnesses we *may* arrive at the real truth, we may form for ourselves an idea of the circumstances of an occurrence and satisfy ourselves concerning it, but the evidence will seldom be true and material; and whoever goes more closely into the matter will not silence his conscience, even after listening to ten witnesses. Evil design and artful deception, mistakes and errors, most of all the closing of the eyes and the belief that what is stated in evidence has really been seen, are

characteristics of so very many witnesses that absolutely
unbiased testimony can hardly be imagined.

At the end of the nineteenth century, Hans Gross wrote that
"A large part of a criminalist's work is nothing more than a battle
against lies. He has to discover the truth and must fight the lie. He
meets the lie at every step. Utterly to vanquish the lie, particularly
in our work, is of course, impossible, and to describe its nature
exhaustively is to write the natural history of mankind."

The trick, of course, is knowing how to recognize a lie when
you hear it.

Involuntary changes in blood pressure are a measure of,
among other things, psychological stress. In 1854 the German
physiologist Karl von Vierordt (1818–1884) developed a device
he called a sphygmograph, which measured instantaneous blood
pressure. In 1863 Etienne-Jules Marey (1830–1904) devised
a portable version of the sphygmograph, which could trace its
readings onto a sheet of paper.

These were further improved upon and then used for the detection of psychological stress by the Italian criminalist Cesare
Lombroso. His apparatus consisted of a glove that would moni-
tor the pulse and changes in blood pressure of the person wearing
it. The glove was in turn connected to a recording pen that made
marks on a revolving drum. In the 1890s Lombroso conducted
experiments in lie detection. He came to his task with a heavy
load of preconceptions, however—he believed that criminals possessed inherited criminal traits and could be recognized by their
appearance. He documented his results, claiming success in using his lie-detection device against real suspects in real criminal
cases, in the 1895 edition of his book *L'Homme Criminel* (*The
Criminal Man*).

The next step in lie detection was taken in 1914 by Vittorio
Benussi, who published the results of research on the differences
in breathing rates between the "honest and dishonest states of

being." He divided the time it took to take a breath by the time it took to exhale. He discovered that the ratio was greater before telling the truth and decreased afterward, but was smaller before telling a lie and increased afterward.

In 1915 the psychologist William Moulton Marston (1893–1947) began researching the effects of lying on systolic blood pressure (the higher number on the blood pressure fraction). In 1917 he wrote a report of his experiments for the *Journal of Experimental Psychology*. By the early 1920s, with the aid of his wife and fellow psychologist Elizabeth Holloway Marston, he believed he had devised a useful mechanism for detecting lies. A Harvard graduate with both a law degree and a Ph.D. in psychology, Marston was a man of many interests. He wrote essays on popular psychology, lived with two women—his wife and his mistress Olive Byrne—ardently supported women's rights, and created the comic book character Wonder Woman, believed to be an amalgam of the two women in his life. Wonder Woman's "lasso of truth" was certainly inspired by the Marston lie detector.

In 1938 Marston wrote a book called *The Lie Detector Test* to support his place in the development of what eventually became the polygraph machine. His publisher sent a copy to J. Edgar Hoover at the FBI, where it was reviewed by Quinn Tamm of the Technical Laboratory. It was not regarded with favor. The reviewer found Marston egotistical and self-aggrandizing. Tamm was particularly displeased with the chapter called "Love and the Lie Detector," in which Marston claims to have settled marital difficulties through the use of the instrument. Tamm commented:

> It is noted that throughout the book the author points out
> that the blood pressure test for the detection of deception in
> the hands of a trained operator is infallible and that once the
> deception has been detected it has been his experience that
> if this is pointed out to the subject he will admit his guilt
> and it will have the psychological effect on him of making

him always in the future tell the truth. This . . . exemplifies the same egotistical ridiculous strain in which this book is written.

Marston's systolic pressure recorder became only the first step toward the device that we think of today as the lie detector.

In 1921 Dr. John A. Larson, working under the direction of Berkeley, California, police chief and famed criminalist August Vollmer, cobbled together a device he called a "cardiopneumo psychograph." It could measure blood pressure, pulse, and breathing rates, but it was large, bulky, and took half an hour to set up. Pens on the device scratched lines on paper that was "smoked" with carbon black and then passed over large recording drums. If a permanent record of the session was required, the paper was shellacked to prevent it from smudging, and it was then stored in cans.

In 1923 a young college student named Leonard Keeler showed Chief Vollmer his plans for an improved lie-detecting device. It was more compact, easier to use, and added galvanic skin response to the measurements it took. "If you build it," Vollmer told him, "we'll try it." Keeler and a couple of his friends built it and dubbed it the "emotograph." It looked, said Vollmer, "like a crazy conglomeration of wires, tubes, and old tomato cans."

According to Ken Alder in his book *The Lie Detectors*, the first success of the emotograph in a murder case involved a bit of highly unorthodox role-playing.

In the early hours of the morning of December 16, 1923, in Los Olivos, California, a small town some forty miles north of Santa Barbara, the shack of itinerant blacksmith John J. McGuire was blown to smithereens. McGuire was asleep inside at the time. The logical suspect, the town's barber and postmaster William H. Downs, who had feuded with McGuire for years, had an unimpeachable alibi—he had spent the night with his family in a hotel in Hollywood.

Williard Kemp, an assistant district attorney of Santa Barbara County, had heard of Keeler's emotograph and decided to try it. The question was how to get Downs to agree to be tested.

At the time the Ku Klux Klan was still very big in rural California, and Kemp was king kleagle of the local chapter. He suggested to Downs that a good way to keep the police off his back would be to join the Klan. And indeed, at that time and in that place, this was probably true. So a fake initiation ceremony was arranged wherein Downs was strapped into Keeler's emotograph and told that he had to tell the truth: "In the solemn secrecy of this room you may answer the questions put to you without fear or favor." Thus Downs took the test but did not tell the truth. Still, when Keeler analyzed the results, he noticed an interesting pattern. When asked if anyone on a list of twenty-five men had been involved in the bombing, Downs had reacted strongly to seven of the names.

Five of the seven were rounded up for questioning. The questioning went on for several days, and the men were hooked up to the emotograph from time to time in order to add to the pressure on them. Finally Harvey Stonebarger, the owner of a local machine shop, broke down and confessed. It had been Downs and he along with Downs's father and a local rancher named William Crawford. (Crawford's was one of the names that Downs had reacted to.) Downs and his father had purchased the explosive, Stonebarger had helped place it, and, after Downs left town, Crawford had lit the fuse with his cigar.

But at the trial Stonebarger retracted his confession, saying it had been obtained by the police using coercive methods. The jury believed him. They discounted the newfangled emotograph and found the men not guilty.

Fred Inbau (1909–1998), a criminologist and law professor at Northwestern University, was a firm believer in the use of lie detectors, but only as an awe-inspiring prop. Inbau joined North-western Law School's Scientific Crime Detection Laboratory in 1933. By the time the lab was taken over by the Chicago Police Department in 1938, he had become its director.

Inbau developed deceit, deception, trickery, and outright lies as interrogation techniques that would be used to trick suspects into confessing. He saw this as an acceptable substitute for the third-degree methods applied in the back room of the police station. These police tactics succeeded in eliciting confessions but, as with other methods of torture, compelled the innocent as well as the guilty to say what the questioner wanted to hear. Inbau's book, *Criminal Interrogation and Confessions*, written with John E. Reid, has been the textbook of choice on the subject since it was first published in 1962. Inbau held, to put it mildly, conservative opinions. He believed that, short of physical persuasion, any method of obtaining a confession was acceptable.

When *Miranda v. Arizona* was decided in favor of Miranda and the Supreme Court mandated that all suspects had to be read their rights—now known as "Miranda rights"—Inbau was in violent opposition and wrote many scholarly articles arguing against it. As Professor Yale Kamisar put it, "One of Professor Inbau's pervasive themes was that, no matter how the officer misbehaved, so long as his or her misconduct did not affect the reliability of the evidence, the evidence should be admissible. It should suffice that the officer was disciplined in a separate proceeding."

Inbau saw the polygraph as a wonderful psychological tool for inducing suspects to confess. And with any luck the investigators would not have to use it at all. As he says:

> There have been innumerable instances of confessions made as a result of a mere proposal to have a suspect submit to a lie-detector test. On many occasions suspects have confessed their guilt while waiting in the laboratory to be tested.
>
> There also are many instances when suspects have confessed

immediately after the operator had adjusted the instrument preparatory to making the test.

He then goes on to reassure the reader that none of these have been false confessions. But he does not say how he knows this.

The sound spectrograph was developed by Bell Telephone Laboratories in 1941 to analyze the frequency of the sound of the human voice as words are spoken. It seemed possible that the characteristics of each person's voice were individual enough that someone could be identified by the graph of his voice. This proved to be so. During World War II voice analysis was used to identify the speakers on German military radio networks. Knowing which radio operator was using which radio helped in keeping track of troop movements and in sorting out the relationships between various units.

In the 1960s, at the urging of the FBI, Bell Labs did further work on voice spectrograms. Bell engineer Laurence Kersta did much of the research and became convinced that "voiceprints," as he called them, could be used for identification with a high degree of accuracy. The word "voiceprint" was dropped, but the technique gained forensic acceptance after a 1972 Michigan State University study that quantified its effectiveness. The study found a false identification rate (in which the analyst picks the wrong person) of only 2 percent, and a false elimination rate (in which the analyst says there is no match when in fact there is) of only 5 percent.

The sound spectrograph was used in 1971 at the behest of a group of reporters who wished to verify that the voice they were listening to over the phone was in fact that of the billionaire recluse Howard Hughes. He was trying to debunk a spurious biography of him that had just been published, and reporters wanted to be sure that it was indeed Hughes doing the debunking. (It was.) The first time most of the public became aware of the technique was in 1972 when Richard Nixon's White House tapes were authenticated.

Since 2002, analysts at the National Security Agency and the Central Intelligence Agency have been busy verifying the voices of Osama bin Laden and other al Qaeda leaders. This is especially true whenever a new tape of their threats is released.

The sound spectrograph is not to be confused with the Voice Stress Evaluator, or Psychological Stress Evaluator (PSE). Purportedly, this device can determine if a person is lying by analyzing the stress patterns in his voice. The efficacy of the PSE is currently in great dispute.

15 : Junk Science

FORENSIC SCIENTISTS are a group of experts from many fields who share an ability to bend the rigor of science to the needs of justice. Whether they are specialists in pathology, anthropology, psychology, chemistry, or serology, all forensic experts have one thing in common—their willingness and ability to use their specialized knowledge to examine and interpret the evidence of a crime.

Studies have shown that when an expert speaks, juries listen. When one side puts an expert on the stand, the other had better respond with an opposing expert. This has led to a proliferation of "hired guns," forensic experts who are adept at tailoring their testimony to aid the prosecution or the defense, depending on which side is paying their fee. Experts from state and local forensic labs are, of course, paid out of public funds—their job is to provide information to the police and prosecutors. This inevitably colors their results and their testimony. One of the suggestions of the 2009 report of the National Research Council on "Strengthening Forensic Science in the United States" is that forensic laboratories become as independent as possible by severing their ties with the police and prosecuting attorneys.

Even beyond this systemic bias, the expert system has occasionally gone wrong, and sometimes horribly so. On the basis of "scientific facts" that are neither scientific nor facts, people have been sent to prison, and some have even been sentenced to death.

Junk science is of two types: pseudo-science, in which a body of knowledge is accepted as science and its conclusions acted upon until a major fallacy becomes evident; and valid science that has been misinterpreted, misstated, or misapplied by practitioners who are uninformed or biased.

A landmark example of pseudo-science is phrenology, a belief system that was widespread in the nineteenth century. It asserted that one could deduce personality and character traits from the shape and contours of the head, a concept with no scientific foundation whatsoever.

The founder of phrenology was the Viennese doctor Franz Joseph Gall. In 1819 he published *The Anatomy and Physiology of the Nervous System in General, and of the Brain in Particular, with Observations upon the Possibility of Ascertaining the Several Intellectual and Moral Dispositions of Man and Animal, by the Configuration of Their Heads*. Gall believed that the shape of the head was an indication of the size of the various "organs" (by which he meant structures) of the brain beneath, and that these organs were responsible for the individual's intellectual and moral capacities. He devised a chart of the head indicating the location of bumps that corresponded to various traits. Among twenty-seven brain organs he identified were intelligence, affection, courage, pride, vanity, guile, benevolence, musicality, a talent for architecture, and the ability to murder.

To this list followers of Gall added gender and racial differences, and whatever else suited their interests. The criminologist Cesare Lombroso found in the precepts of phrenology the ability to predict criminality and to differentiate southern Italians from northern Italians. At the end of the nineteenth century phrenology was used to assess the vocational aptitudes of children, to appraise marriage prospects, and to scrutinize job applicants.

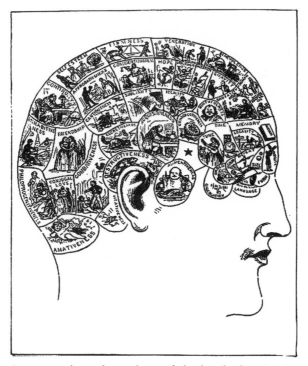

An 1883 phrenology chart of the head, showing its various brain organs.

When Lord William Russell was found murdered in his bedroom on May 6, 1840, suspicion focused on François Courvoisier, his newly hired valet. Courvoisier had planted a few diversionary clues in an attempt to suggest that someone had broken in from the outside. Nonetheless he was tried, sentenced to death, and, on July 6, 1840, hanged. A cast was taken of Courvoisier's head, and Dr. John Elliotson (1791–1868), a founding member of the Phrenological Society and author of *Surgical Operations in the Mesmeric State without Pain* (1843), studied the cast to see what phrenology might reveal about François Courvoisier. He reported:

> In the sides and back and in the posterior portions of the crown reside the dispositions which are as powerful in

brutes as in man—love, sexual, parental, and friendly; the disposition to resist, to do violence, to act with cunning, to possess, to construct; self-estimation; love of notice; cautiousness; and firmness.

The organs of the whole of these are very large, with the exception of those of Parental and Friendly Love, which are but mean or ordinary. . . . The part . . . situated just before the organ of the Disposition to do Violence is also very large. . . .

[This] must suggest the most serious reflections to every thinking and humane person upon the improvements necessary in education, in the views and arrangements of society, and in punishments and prison discipline.

A forensic odontologist identifies deceased victims by their dental records. Since the time of the Roman Empire, dentists have been called upon to perform this service. Their services are still in demand today, particularly at scenes of mass violence or large-scale tragedy. Teeth are often the only parts of the body that survive in recognizable form after a fire. As recounted in Chapter 13, it was the survival of Dr. Parkman's teeth when much of his body was consumed in a furnace that put the noose around the neck of the first Harvard professor to be convicted of murder.

Some forensic odontologists will also attempt to link bite marks left on a victim to the teeth of a particular suspect. The use of this generally unreliable method stems from its successful application in two notorious cases. In 1967 Gordon Hay killed a schoolgirl in Scotland and left a distinctive bite mark (due to malformed teeth) on her left breast. Serial killer Ted Bundy, believed to have killed more than forty women between 1973 and 1978, left bite marks on the left buttock of Lisa Levy, one of his Florida State University victims. Forensic odontologist Dr. Richard Souviron used an acetate overlay of Bundy's front teeth to demonstrate how they fit exactly over a photograph of the bite marks.

On May 23, 1991, a farmhouse in upstate New York was set afire in order to cover up the murder of its occupant, a social worker. She had been beaten, strangled, bitten, and stabbed to death. Police collected a bloody nightshirt at the scene and swabbed the bite marks for saliva.

They subsequently arrested Roy Brown, a man who had just served a short jail sentence for making threatening phone calls to the director of the social service agency where the victim worked. A year before, this same agency had removed Brown's daughter from his custody and placed her in residential care. Still, the victim had not been involved in Brown's case, and there is no evidence that Brown had known her.

Things went rapidly downhill for Brown. A snitch who had met Brown in jail claimed that Brown had called him and confessed to the killing. A local dentist examined the bite marks and concluded that they had been made by Brown's teeth. At the trial the defense's expert pointed out that the bite marks showed the imprint of six teeth, whereas Brown was missing two front teeth. The prosecution's dentist testified that Brown could have twisted the victim's skin in such a way that the marks would appear to show the two missing teeth. Why he should do this was not discussed. The jury deliberated for five hours before finding Brown guilty.

Brown continued to fight to prove his innocence. When a fire destroyed all his court records, he filed for copies of them under the Freedom of Information Act. Among the documents he received were some that had not been previously disclosed to the defense. These included the report of another prosecution odontologist who had found that the bite marks excluded Brown, and evidence that there was another suspect, Barry Bench, a man with a grudge against the victim. Bench was reported to have been "acting strangely" around the time of the murder. Brown asked for DNA testing on the saliva samples taken from the bite marks but was told that the samples had been used up.

Then, when Brown wrote to Bench and informed him that he would be testing the saliva, Bench promptly committed suicide by stepping in front of a train.

In 2005 the Innocence Project took on Brown's case and discovered saliva exemplars on the nightshirt found at the scene. The DNA test showed that they did not come from Brown. Comparison with DNA from Bench's daughter showed a 50 percent match, exactly what one would expect to find if Bench were the donor.

On Tuesday, January 23, 2007, Roy Brown was released from prison. One month later the district attorney dropped all charges against him.

In 1999 a member of the American Board of Forensic Odontology conducted a study showing that bite-mark evidence was wrong 63 percent of the time. And yet forensic dentists continue to provide sworn testimony that *this* bite mark was caused by *this* defendant. And juries continue to take them seriously.

"If you say that this bite fits this person and nobody else in the world, and if you use the bite mark as the only piece of physical evidence linking an attacker to his victim, that's not science, that's junk!" says Dr. Richard Souviron, chief forensic odontologist at the Miami-Dade medical examiner's office. According to Souviron and many other experts, using bite-mark evidence to do anything other than exclude or include a person as one possible donor among many is a miscarriage of justice.

In the section on handwriting of his landmark book *Criminal Investigation*, Hans Gross writes, "The most important thing an Investigating Officer can extract from a writing is, in every case, the character of an individual." In this he was completely mistaken. Whether a person is good or bad, sharp or dull, easygoing or driven cannot be reliably assessed from his or her handwriting. It may be possible to judge the age of the writer, but a number of neurological problems can masquerade as the palsy of age.

Although completely innocent of the charges against him, Captain Albert Dreyfus of the French army was convicted of treason in December 1894 on evidence manufactured by the Deuxieme Bureau, the army's intelligence service, in order to cover up the army's own ineptitude. Dreyfus was selected for this honor principally because he was a Jew. France, and in particular the French army, was at that time suffering an epidemic of anti-Semitism. After five years on Devil's Island and another six during which the army knew him to be innocent but refused to admit it, Dreyfus's conviction was quashed and he was reinstated in the army with the rank of major.

The novelist Emile Zola wrote a now-famous editorial, "J'accuse," accusing the government of complicity in Dreyfus's unjust court-martial. Zola was himself prosecuted for defaming the government and sentenced to a year in prison, though the sentence was quickly reversed.

The Dreyfus Affair, as it became known throughout the civilized world, nearly brought down two successive French governments, caused a major shakeup in the army, and engendered a complete loss of faith in the integrity of the intelligence department.

The affair began in September 1894 when Agent Auguste of the Deuxieme Bureau intercepted a letter that had been left at the German embassy for Lieutenant Colonel Maximilien von Schwartzkoppen, the German military attaché:

Having no indication that you wish to see me, I am nevertheless forwarding to you, Sir, several interesting items of information

    1. A note on the hydraulic brake of the 120 and the manner in which that part has performed;

2. A note on covering troops (several modifications will be effected by the new plan);

3. A note on a modification of Artillery formations;

4. A note pertaining to Madagascar;

5. The *Sketch for a Firing Manual* for the country artillery (March 14, 1894).

This last document is extremely difficult to procure and I am able to have it at my disposal for only a very few days. The Ministry of War has distributed a fixed number of copies to the regiments, and the regiments are responsible for them. Every officer holding a copy is to return it after maneuvers. If you would then take from it what interests you and keep it at my disposal thereafter, I will take it. Unless you want me to have it copied in extenso and send you the copy.

I am off to maneuvers.

The letter allowed of no other interpretation than that a high-ranking French officer was spying for the Germans. A copy of the letter was taken to General Auguste Mercier, the minister of war, who took direct and forceful action. "The circle of inquiry is small," he is reported to have informed his chief of staff, "and is limited to the General Staff. Search. Find."

The letter, which would become known in France as *le bordereau* (the memorandum, as though there had never been another), had been torn into six pieces before being reassembled by the section of statistics. This was unusual for material from von Schwartzkoppen's wastebasket—his usual practice was to tear letters into very small pieces. He later claimed never to have seen the *bordereau* at all.

The contents of the *bordereau* were analyzed by Major Hubert Henry of the section of statistics. Because of some of the letter's references (the new plan and the secret artillery manual), he decided that the writer was indeed an officer with the General Staff, and further that he was an artillery officer.

So the authorities began looking for an artillery officer who had recently been or was still attached to the General Staff. When they went over the list of possibilities, the name of Captain Albert Dreyfus appeared. He was from Alsace, was he not? And Alsace was practically German. And he was a *Jew*. They found a sample of his handwriting and compared it to that on the *bordereau*. It was similar! (This would hardy seem surprising—Dreyfus wrote in the same slanting, precise hand that was taught at many schools in those days.)

A handwriting expert, Alfred Gobert, was given the *bordereau* and samples of Dreyfus's handwriting. After a long and careful examination, he concluded that the *bordereau* had not been written by Dreyfus. He was then removed from the case and Alphonse Bertillon, the noted criminologist, was called in. In a matter of hours he concluded that Dreyfus had written the document. To bolster his opinion, two army colonels, Favre and d'Abboville, were made *ex tempore* graphologists. They too could see clearly that Dreyfus had written the *bordereau*. Their joint testimony would more than overbalance that of Gobert, who was not even an army officer.

Other evidence against Dreyfus was now assembled, much of it forged. Added to the testimony of Bertillon, who was justly famous for his development of anthropometrics, it would more than suffice. Bertillon was called to testify at Dreyfus's trial, and his assertion that the handwriting was Dreyfus's, along with the flamboyant testimony of Major Henry, was damning. Henry capped his testimony by pointing dramatically to a painting of the crucified Christ and declaiming, "I swear to it!" The conviction was unanimous. Dreyfus was sent to the penal colony on Devil's Island.

Later that year Lieutenant Colonel Georges Picquard, head of the Deuxieme Bureau, was shown an intercepted letter that had been written by Lieutenant Colonel Schwartzkoppen, the German attaché, to Major Ferdinand Esterhazy of the French army.

Picquard investigated and had samples of Esterhazy's handwriting analyzed. The fact was self-evident: it was Esterhazy who had written the *bordereau!*

When Picquard brought this new information to the attention of his superiors, he was promptly removed as head of the Deuxieme Bureau and sent to a new post in North Africa. Major Henry was installed in his place. To keep the lid on the situation and to keep Dreyfus safely on Devil's Island, Henry immediately concocted an outrageous forgery. He pasted together parts of a real letter from Schwartzkoppen with his new fabrication, in a manner incriminating to Dreyfus.

In due course, and only because it proved to be unavoidable, Esterhazy was charged with the crime and court-martialed. Picquard was brought back to testify. When Esterhazy was exonerated, Picquard was arrested.

Meanwhile it was becoming increasingly clear to impartial observers that Dreyfus had been unjustly convicted and that, while he suffered on Devil's Island, the true villain walked the streets of Paris a free and even celebrated man. As more facts were uncovered, the ranks of the Dreyfusards (those who believed in Dreyfus's innocence), grew. Equally, the regiments of the anti-Dreyfusards swelled and their anti-Semitic hatred of Dreyfus increased. Some high-ranking French army officers were by now aware of their mistake but chose to remain silent. The honor of the French army, indeed of France herself, was at stake—General de Boisdeffre of the General Staff swore that he had seen a document that *proved* Dreyfus's guilt.

When Esterhazy was acquitted, Emile Zola felt compelled to protest. In the newspaper *L'Aurore*, he published the manifesto he entitled "J'Accuse." It read in part:

> I accuse General Billot of having had in his hands definite evidence of the innocence of Dreyfus and of having stifled it, of being guilty of an outrage against humanity and an

outrage against justice for a political end and in order to save the compromised General Staff. . . .

I accuse the three handwriting experts [in the Esterhazy trial], Mssrs. Belhomme, Varinard, and Couard, of having composed deceitful and fraudulent reports, unless a medical examination declares them to be stricken with an impairment of vision or judgment. . . .

Finally I accuse the first Court Martial of having violated the law in convicting a defendant on the basis of a document kept secret, and I accuse the second Court Martial of having covered up that illegality on command by committing in turn the juridical crime of knowingly acquitting a guilty man.

The authorities' response was to arrest Zola and try him for insulting the government. Bertillon was once again called to testify. This time his description of the manner in which the handwriting on the *bordereau* had been "forged" by Dreyfus showed that he completely lacked credibility.

In order to guide his handwriting, the writer of the bordereau made use of a kind of transparent template inserted at every line beneath the tracing paper of the bordereau. The template consists of a double chain: the first chain is composed by the word "*intérêt*" traced end to end indefinitely and meshing with each other; that is written in such a manner that the initial *I* merges with the final *t* preceding it; the second chain is identical to the first, but displaced 1.25 mm. To the left.

The transparent template is traced from the word *intérêt*, which concludes a letter found in Dreyfus's blotter. The word itself is not written naturally, but constructed geometrically.

Zola was found guilty and sentenced to a year in prison. He contemplated serving the term as an expression of his contempt for the current government, but his editors convinced him that freedom was the better part of valor. He fled to London.

On August 13, as he carefully examined all the documents in the case at the direction of the minister of war, Captain Louis Cuignet made an astounding discovery. The letter that Major Henry had "found" was pasted together from two different paper stocks. The thin, colored lines that ran through the two types of paper did not match. Not only that, but the lines were of entirely different colors. It was clear that Henry had forged the document.

General de Boisdeffre, who had sworn in public to the authenticity of the document, promptly resigned. Esterhazy was discharged from the army. Henry was taken to prison for further questioning. The next day, while left alone in his cell, he slit his throat with a straight razor.

In 1899, on the basis of the new evidence and the changed circumstances, Alfred Dreyfus was returned to France from Devil's Island to stand trial once again for treason. And once again he was found guilty.

The President of France, deciding that enough was enough, granted Dreyfus a pardon. But still he was not vindicated. In 1904 the court of appeal agreed to rehear the case, and in 1906 the conviction was thrown out. Dreyfus was reinstated in the army as a captain. He retired in 1907 but was called up for service again during World War I.

On March 1, 1932, the infant son of aviation hero Charles Lindbergh was kidnapped from the Lindberghs' home in East Amwell, New Jersey. The kidnapper left behind a ransom note:

Dear Sir!
    Have 50.000$ redy 25.000$ in 20$ bills 15000$ in 10$ bills and 10.000$ in 5$ bills. After 2–4 days we will inform you were to deliver the mony.

The ransom note from the Lindbergh kidnapping. Experts agreed it had probably been written by a semi-literate German immigrant.

We warn you for making anyding public or for notify the polise the child is in gut care.

Indication for all letters are singnature and 3 holes.

In the bottom right-hand corner of the note was the "singnature"—a pair of inch-wide, interlocking blue circles with the overlapping space colored red. In it, there were three holes that might have been pushed through with a pencil. More letters followed, all similarly misspelled, and all marked with the same identifying symbol.

The letters were given to several experts for analysis. On the basis of the misspellings and the phraseology in the letters, the experts agreed that the writer was in all probability a semi-literate German immigrant. Handwriting analyst Albert S. Osborn contrived a paragraph that the police might place before a suspect and ask him to copy. It contained several of the words that the kidnapper had misspelled, such as money (mony).

The ransom was paid, but the child was not returned. A few weeks later the severely decomposed body of a small child was found in the woods about four miles from the Lindbergh house.

In 1934 a man with a German accent stopped at a gas station and bought 98 cents' worth of gas, paying for it with a ten-dollar gold certificate. Gold certificates, which were paper bills directly backed by gold, had been withdrawn from circulation in 1933. The attendant, fearing that the bill might be counterfeit, noted the license plate of the car. The certificate turned out to be from the Lindbergh ransom. The car belonged to Bruno Richard Hauptmann.

Hauptmann was asked to provide samples of his handwriting. Albert Osborn and his son, Albert D. Osborn, compared these samples with the writing on the ransom notes. To the Osborns they did not look similar. The police instructed Hauptmann to write more notes, telling him just how to spell certain words. These too were not close enough to convince the Osborns. There were too many dissimilarities between Hauptmann's writing and that of the ransom notes, and to many discrepancies among the writing samples themselves.

When a search of Hauptmann's garage turned up more than $14,000 in additional ransom money, he was arrested. He claimed that Isador Fisch, a friend who had returned to Germany, had left the package of money with him. He was not believed. And the discovery of the money convinced the Osborns that the writing on the ransom letters was Hauptmann's after all.

Eighteen months later at Hauptmann's trial the ransom notes were introduced into evidence along with the samples of Hauptmann's writing. No one bothered to tell the jury that Hauptmann had actually been given the notes to copy from, and had been made to practice for hours until his handwriting matched that on the notes.

The handwriting exemplars and the ladder that had been used in the kidnapping were the primary pieces of evidence against Hauptmann. The state maintained that it had proved that the ladder had been made from lumber in Hauptmann's garage. Today there are people who feel that the ladder evidence was at best exaggerated.

Hauptmann was executed in New Jersey's electric chair on April 30, 1936.

As chief of serology at the West Virginia State Police Crime Laboratory from 1980 to 1989, Fred Zain testified as an expert witness in hundreds of criminal cases. His manner on the stand was imposing and his command of facts impressive—when he testified for the prosecution, the prosecution tended to win. He resigned in 1989 to take a more lucrative job as chief of physical evidence for the medical examiner in Bexar County, Texas.

Although it would not catch up with him for a few years, Zain's downfall began in 1987 when he testified at the rape trial of Glen Woodall. According to Zain, analysis of the semen found at the rape scene showed that "[t]he assailant's blood types . . . were identical to Mr. Woodall's." Woodall was convicted and sentenced to serve between 203 and 335 years in prison. Five years later, when Woodall and his attorney were finally able to get DNA testing of the semen, it was shown conclusively not to be his. Woodall sued West Virginia for false imprisonment and settled for $1 million. When the West Virginia Supreme Court decided to look into Zain's testimony, it asked retired circuit judge James O. Holliday to conduct an investigation of Zain's tenure with the West Virginia Crime Laboratory.

Five months later Holliday issued a report in which he concluded that throughout his tenure at the lab, Fred Zain had been consistently guilty of:

[1] Overstating the strength of results; [2] overstating the frequency of genetic matches on individual pieces of evidence; [3] misreporting the frequency of matches on multiple pieces of evidence; [4] reporting that multiple items had been tested, when only a single item had been tested; [5] reporting inconclusive results as conclusive; [6] repeatedly altering laboratory records; [7] grouping results to create the erroneous impression that genetic markers had been obtained from all samples tested; [8] failing to report conflicting results; [9] implying a match with a suspect when testing supported only a match with the victim; and [10] reporting scientifically impossible or improbable results.

Trooper Zain's misconduct thus significantly tainted his testimony in numerous criminal prosecutions. In this regard, the Holliday report stated: "It is believed that, as a matter of law, any testimonial or documentary evidence offered by Zain at any time in any criminal prosecution should be deemed invalid, unreliable, and inadmissible in determining whether to award a new trial in any subsequent habeas corpus proceeding."

Kenneth Blake, who had been director of the West Virginia State Police's Criminal Identification Bureau, was later asked about the complaints that had been lodged against Zain by others at the lab, in which they alleged that he had filed false reports. At the time, Blake had dismissed the allegations as an office squabble. "They didn't like Zain, and Zain didn't like them," Blake had explained. "But we never had any complaints from prosecutors, defense attorneys, or investigators."

In 1995 Baltimore police sergeant James A. Kulbicki was found guilty of the first-degree murder of twenty-two-year-old Gina

Marie Neuslein and sentenced to what criminal attorneys call LWOP—life in prison without the possibility of parole. Twelve years later the so-called scientific evidence that formed the basis of his conviction had fallen to pieces.

Neuslein had been having an affair with Kulbicki, a married man, and had just had his child. She was planning to sue him for child support. Joseph Kopera, Maryland's top firearms expert, testified that Kulbicki's gun had been recently cleaned and that the murder bullet was "consistent in size" with the ones used by Kulbicki's gun. An FBI expert testified further that the lead fragments found in the victim—all that remained of the murder bullet after it entered Gina Neuslein's skull—were identical in composition with the lead in Kulbicki's bullets. Against this expert testimony were defense witnesses who had seen Kulbicki running errands at the time of the murder. They were not believed.

Then, in 2005 the FBI announced that it would no longer do bullet lead analysis, as the conclusions reached by that technique were unreliable. And in 2007 Maryland state public defenders, working with the Innocence Project, discovered that ballistics expert Joseph Kopera had lied about his credentials, falsely claiming degrees from two universities and forging at least one document to back up his claim. The day after this discovery, Kopera committed suicide.

When they examined the notes that Kopera had made when he tested Kulbicki's gun, investigators found that they flatly contradicted the testimony he had given. He had found in fact that Kulbicki's gun could not have fired the bullet that killed Neuslein.

"If this could happen to my client, who was a cop who worked within the justice system, what does it say about defendants who know far less about the process and may have far fewer resources to uncover evidence of their innocence that may have been withheld by the prosecution or their scientific experts?" asked Kulbicki's attorney, Suzanne K. Drouet, a former Justice Department lawyer.

At this time Kulbicki's request for a new trial is still pending.

No doubt most forensic analysts are honest, dedicated, and competent, follow the evidence impartially, and do not manipulate it to suit their own biases. Those misguided few who knowingly doctor the evidence honestly believe they are doing it in a good cause—helping the prosecution put away a "bad guy."

Even when we add these to the incompetent, we still have a very small number of criminalists who taint the profession. But these few can do great damage to the criminal justice system.

To address this problem the National Research Institute suggests that all forensic science professionals be certified, with standards for certification established by the National Institute of Forensic Science based on the recognized standards for each discipline. It recommends further that "No person (public or private) should be allowed to practice in a forensic science discipline or testify as a forensic science professional without certification."

This will help eliminate those with fraudulent credentials, and rigorous testing and recertification every few years will keep the true professionals up to date.

Forensic evidence, properly analyzed, is more reliable than eyewitness testimony, more reliable than the testimony of supposed accomplices, more reliable even than the freely given confession of a suspect. With each passing decade, more scientific techniques become available to the forensic technician. These should be made available to the prosecution and the defense alike. Further, all physical evidence should be carefully preserved in the event that tests we cannot now even imagine may become available in the future to affirm or refute the decisions of today's courts. The judicial doctrine of finality should not be applied to those on death row or to those spending the rest of their lives behind bars.

16 : A Double-Breasted Suit

THE PSYCHOLOGICAL PROFILER is the new magician of forensic science. And as with magic, profiling provokes a variety of responses from observers and practitioners alike. The basic notion is simple. If you can think like a criminal, you can predict his actions. Or if you have many examples of his actions, you can form a clear image of the sort of person he is. Both of these notions have been tried, and on occasion they have worked exceedingly well. And sometimes they have not.

Probably the most infamous murderer of all times is the man known only as "Jack the Ripper." He terrorized London for a few months in 1888, killing five women that we know of and possibly at least two more. The victims, all prostitutes working in the Whitechapel district of London's East End, were killed quickly and brutally, some on street corners where anyone might walk by at any second—in one case it seems he was interrupted before he had finished. The victims were horribly mutilated, and with each killing the mutilation increased—the murderer took body parts away with him when he left the scene. After one killing he mailed half the victim's kidney to the head of a Whitechapel vigilante organization, claiming in his letter that he had fried and eaten the other half. He ended his letter with the taunt, "Catch me if you can."

Another victim of Jack the Ripper in London's East End.

Police surgeon Dr. Thomas Bond, who had assisted in the autopsy of Mary Kelly, one of the victims, was asked by investigators for a medical analysis of the wounds. In notes dated November 10, 1888, Bond provided what today would be called a
profile of the probable killer.

The killings, Bond asserted, had been committed by one man
working alone. He was probably middle-aged, quiet, and inof-
fensive looking but nonetheless physically strong, composed,
and daring. He dressed neatly and probably wore a cloak to hide
any bloodstains resulting from his efforts. He would be a loner,

without any real occupation, eccentric, and mentally unstable. He might suffer from Satyriasis (today we call it hypersexuality), and his acquaintances would be aware that he "was not in his right mind." Bond was quite certain that the Ripper had a sixth victim, a prostitute named Alice McKenzie, though her wounds were unlike the others'. He thought that offering a reward for information might nudge someone's memory or conscience.

Since the Ripper was never caught, the accuracy of Dr. Bond's assertions cannot be known. Much has been written about the Ripper case and many theories cited as to who the Ripper was. None have been wholly satisfactory.

Before the days of official police profiling, there was the case of George Metesky (1904–1994), New York City's "mad bomber." He began his bombing before World War II, leaving his first bomb in a toolbox on a windowsill in a Consolidated Edison facility on West Sixty-fourth Street. That was on November 16, 1940. The bomb didn't go off, and the bomb squad found a note wrapped around it: "CON EDISON CROOKS, THIS IS FOR YOU." The presence of the note was puzzling, for if the bomb had gone off it would have been destroyed.

The unsuccessful bomber would seem to be a disgruntled employee or former employee of Con Edison. But Con Edison was vast. Except for a miniscule corner of Queens, it served all of New York City and most of Westchester County. And it was an amalgam of more than twenty earlier companies, some added only recently. The records of the employees of all these companies were scattered and poorly kept.

The next bomb also failed to go off. It was found almost a year after the first one, wrapped in a sock and lying on Nineteenth Street near a Con Edison office. Three months later Pearl Harbor was attacked, and New York, along with the rest of the country, had more important things to do than hunt down a maker of dud bombs. The bomber agreed, sending a letter to the police:

> I WILL MAKE NO MORE BOMB UNITS FOR THE DURATION OF THE
> WAR—MY PATRIOTIC FEELINGS HAVE MADE ME DECIDE THIS—
> LATER I WILL BRING THE CON EDISON TO JUSTICE—THEY WILL
> PAY FOR THEIR DASTARDLY DEEDS . . . F.P.

And F.P. was true to his word. Apart from a few crank letters and postcards sent to the police, the newspapers, Con Edison, and random strangers, he was not heard from again until 1951, ten years later.

Over the years his bomb-making skills had improved, and his choice of targets became more grandiose. The first bomb of this new batch, and the first to actually detonate, was placed on the lower level of Grand Central Station. Luckily, no one was hurt in the explosion.

Evidently miffed that his bombs were getting so little publicity (the police were urging the newspapers not to play up the incidents), F.P. sent another letter, this time to the *New York Herald Tribune*:

> HAVE YOU NOTICED THE BOMBS IN YOUR CITY—IF YOU ARE
> WORRIED, I AM SORRY—AND ALSO IF ANYONE IS INJURED. BUT
> IT CANNOT BE HELPED—FOR JUSTICE WILL BE SERVED. I AM NOT
> WELL, AND FOR THIS I WILL MAKE THE CON EDISON SORRY—YES,
> THEY WILL REGRET THEIR DASTARDLY DEEDS—I WILL BRING
> THEM BEFORE THE BAR OF JUSTICE—PUBLIC OPINION WILL
> CONDEMN THEM—FOR BEWARE, I WILL PLACE MORE UNITS
> UNDER THEATER SEATS IN THE NEAR FUTURE. F.P.

He went on to place more than thirty bombs over the next six years. His targets included Radio City Music Hall, the Capitol Theater, the Roxy Theater, the Paramount Theater, the New York Public Library, and Macy's department store. In 1954 a second bomb at Radio City went off during a showing of Bing Crosby's movie *White Christmas*, injuring four people. A bomb exploded in a toilet bowl in a men's room at Pennsylvania Station, seriously injuring the seventy-four-year-old attendant who was trying to plunge out a bowl at the time.

While the bombs had not yet killed anyone, the bombings had injured quite a few people and were causing near-panic in the city. And the police were spending too much time following dead-end leads, searching for unexploded bombs, and attending to copycats and pranksters.

By 1956, after F.P. had planted his bombs for sixteen years, the authorities had neither caught him nor even developed any promising leads. This led the New York City police to call on the services of psychoanalyst James A. Brussel to create a personality profile of the mad bomber. After reviewing all the case material, Dr. Brussel started with the obvious:

—The bomber was paranoid.

—The bomber had a specific grievance against Con Edison.

—The bomber was male (well over 90 percent of bombers are male).

—The bomber was around fifty years old (paranoia classically peaks at thirty-five, and this had been going on for fifteen years).

—The bomber was a meticulous worker (the construction of the bombs showed that).

Then Dr. Brussel proceeded into the unknown:

—The bomber was not a native-born American. His use of language on the notes—"dastardly deeds," "the" Con Edison—demonstrated this.

—The bomber probably had a high school education, but probably little or no college. This deduction was also derived from the language of the notes.

—The bomber was of Slavic origin and probably Roman Catholic. Eastern Europeans were known to use bombs, and most Slavs are Catholic.

—The bomber lived in Connecticut rather than New York. Many of the letters had been mailed from Westchester County, halfway between New York and Connecticut. There was a large Eastern European population in Connecticut.

Dr. Brussel also concluded that the bomber had an Oedipal complex. Therefore he was probably unmarried and might live with one or more female relatives, but not his mother.

Dr. Brussel threw in one last observation, "When you catch him," he said, "and I have no doubt you will, he'll be wearing a double-breasted suit." Before the detectives could say anything he added, "And the suit will be buttoned!"

Then Dr. Brussel gave the detectives a piece of advice that went against their standard practices and all their police instincts: publicize these findings. Someone would come in or would have seen something useful. There was even the possibility that the bomber would see something he objected to in Dr. Brussel's profile and call to complain.

So the police released Dr. Brussel's profile along with other information about the bombings, holding back only a few details.

On Christmas Day 1956, acting on Dr. Brussel's suggestion, New York City newspapers published a version of his profile of the bomber. The next day the *New York Journal-American* printed an open letter to the bomber urging him to turn himself in. The paper promised to publish his grievances and to see that he got a fair trial. F.P. answered, listing all the places where he had planted bombs in the past year and saying, "My days on earth are numbered—most of my adult life has been spent in bed—my one consolation is—that I can strike back—even from my grave—for the dastardly acts against me."

The *Journal-American* printed an edited version of his reply and followed it with a request that he tell its readers what his grievances were. He replied. He had been injured, he said, at a Con Edison plant, and had to pay his own medical bills; Con Edison had even blocked his worker's compensation case. He went on:

> When a motorist injures a dog—he must report it—not so with an injured workman—he rates less than a dog—I tried to get my story to the press—I tried hundreds of others—I typed tens of thousands of words (about 800,000)—nobody cared . . . I determined to make these dastardly acts known—I have had plenty of time to think—I decided on bombs.

He then sent a third letter, in which he revealed the date of his injury—September 5, 1931. Alice Kelly, a Con Edison clerk who had been going through old files, found letters from a George Metesky in a batch of denied workman's compensation forms from 1931. She noted that words in his letters were similar to the ones in the bomber's published replies. These included the phrase "dastardly deeds."

New York police went to Metesky's house in Waterbury, Connecticut, a little before midnight on Monday, January 21, 1957. They were accompanied by several local policemen and carried a search warrant.

Metesky answered the door in his pajamas. "I know why you fellows are here," he told them. "You think I'm the Mad Bomber." The detectives asked him what F.P. stood for, and he told them, "Fair Play."

The detectives told him that they were taking him to the Waterbury police station, so he went upstairs to get dressed. When he came back down he was wearing a double-breasted suit, and it was buttoned.

The author and pop psychologist Malcolm Gladwell does not think that profiling is a science at all. With regard to Dr. Brussel and the Metesky case, he wrote:

> Brussel did not really understand the mind of the Mad Bomber. He seems to have understood only that, if you make a great number of predictions, the ones that were wrong will soon be forgotten, and the ones that turn out to be true will make you famous. The Hedunit is not a triumph of forensic analysis. It's a party trick.

But a double-breasted suit? Buttoned?

In 1943 the OSS (Office of Strategic Services), a wartime espionage service and precursor to the CIA, asked Dr. Walter C. Langer to develop a behavioral and psychological analysis of Adolf Hitler. They wanted to be able anticipate Hitler's reactions to various war scenarios.

Langer used a number of sources to build up his profile—Hitler's autobiography, *Mein Kampf* (*My Struggle*), his speeches, and interviews with people who had known him. In the resulting 135-page profile, Langer noted that Hitler was meticulous, conventional, and prudish about his appearance and body. He was not shy about describing himself, however. When Germany reoccupied the Rhineland in 1936, he said, "I follow my course with the precision and security of a sleepwalker." This is hardly the description of a man in conscious control of his actions. And his estimation of himself was in little doubt: "Do you realize that you are in the presence of the greatest German of all time?" and "I cannot be mistaken. What I do and say is historical." Langer predicted that in the event of the military collapse of Germany, Hitler would commit suicide. In 1945, hiding in his bunker in Berlin as the Russian army reached the outskirts of the city, Hitler shot himself.

On December 29, 1977, as Ambrose Griffin of Sacramento, California, headed from his car into his house with two sacks of groceries, he suddenly fell over. His wife thought he was having a heart attack and rushed him to the emergency room, where he died. The cause of death turned out to be wounds caused by two .22-caliber bullets. His wife then remembered hearing two popping sounds just as her husband collapsed.

The following day a news crew found two spent shell casings on the street outside Griffin's house. Later that day a twelve-year-old boy reported that he had been shot at while riding his bicycle the day before. He said that the assailant drove a brown Pontiac Trans Am—twelve-year-old boys are good at identifying cars—had brown hair, and was probably in his twenties.

A woman living two blocks away reported she had been shot at in her house two days earlier. A .22 slug that matched the ones taken from Griffin was found in her kitchen wall.

Almost a month later, on January 23, 1978, twenty-two-year-old Theresa Wallin was savagely attacked as she took out the garbage. A bullet went through her hand, up through her arm, and out her elbow. A second bullet penetrated the top of her head. After she fell to the ground, the killer shot her again in the head, dragged her lifeless body inside and into her bedroom, and furiously assaulted it with a knife. She was stabbed repeatedly and cut open. Her spleen, kidneys, and intestines were pulled out. Then her kidneys were placed back inside her body. Her left nipple was excised. An empty yogurt container stained with blood was found near her body, as though someone had used it to gather and drink her blood. Bloody footprints were found in several rooms.

When Theresa's husband came home after work and found his wife, who was three months pregnant, he ran from the house and could not stop screaming.

Sacramento police feared they had a serial killer on their hands and asked the FBI for help. Profiler Robert Ressler gathered the scant information that was known about the killer and suggested they look for a

> White male, aged 25–27; thin, undernourished appearance. Residence will be extremely slovenly and unkempt, and evidence of the crime will be found at the residence. History of mental illness, and will have been involved in the use of drugs. Will be a loner who does not associate with either males or females, and will probably spend a great deal of time in his own home, where he lives alone. Unemployed. Probably receives some sort of disability money. If residing with anyone, it would be with his parents; however this is unlikely. No prior military record; high school or college dropout. Probably suffering from one or more forms of paranoid psychosis.

There are two kinds of serial killers—organized killers who plan everything methodically, and disorganized killers who plan nothing, act entirely on impulse, and scatter clues around their crime scenes. This one was clearly a disorganized killer, but clues like shoe prints are useful only when you have shoes to compare them to.

Four days later, on January 27, the neighbors of thirty-eight-year-old Evelyn Miroth grew worried when a little girl who had been sent to Miroth's house on an errand reported that although she had seen some movement in the house, no one answered the bell. One of the neighbors entered the house, saw Evelyn's fifty-one-year-old friend Dan Meredith lying dead in the hall, and immediately called the police. The first officer on the scene found Meredith with a gunshot wound to the head. He moved further into the house and found Evelyn in the bedroom. She was naked and shot through the head. She had been sodomized and then cut up in a fashion similar to that used on the body of Theresa Wallin. On the far side of the bed lay Evelyn's six-year-old son Jason, shot twice through the head.

Evelyn had been baby-sitting for a friend, and now the infant was missing. A bullet hole was found in a pillow where the child had evidently been sleeping.

Dan Meredith's red station wagon was taken by the killer and found parked a few blocks away with the keys still in it.

The police surmised that the killer lived within walking distance of where he had left the car and within walking distance of his victims—if he had stolen a car he had probably walked to the crime scene. A woman had told them of seeing a wild-eyed young man driving a red station wagon. Based on this description, they went around the neighborhood with a sketch of their suspect.

When Nancy Holden saw the sketch, it rang a bell. It looked to her like a boy she had gone to high school with. She would not have recognized him at all except that she had recently run into him in a shopping mall and thought that he had acted weirdly. His name was Rick Chase—Richard Trenton Chase.

Two officers went to Chase's apartment, only a few hundred yards from where he had left the station wagon. Although they could hear him moving around inside, he refused to come to the door. They left but staked out the apartment and waited. Chase came out a while later, carrying a cardboard box and heading for his car. When the detectives tried to grab him, he threw the box at them and ran. They chased him and, after a violent struggle, subdued and handcuffed him.

They found a loaded .22-caliber pistol in his belt and a pair of latex gloves and Dan Meredith's wallet in his pockets. In the cardboard box were bloodstained papers and rags.

The detectives took Chase to the police station, then, as he was being questioned, went back to search his apartment. The place was covered with bloodstains and strewn with feces. In the refrigerator were dishes holding human internal organs including brain tissue. There were three dog collars but no dogs. They later found out that Chase once bought dogs in order to kill them. On a calendar he had written "TODAY" across the dates of the Wallin and Miroth killings, with forty-four such dates yet to come.

The predictions of Robert Ressler, the FBI profiler, were eerily accurate. He accurately estimated Chase's age. He was correct about Chase's physical appearance, the appearance of his apartment, and the fact that mementos of the crimes would be found there. Chase indeed had an extensive history of mental illness, but given the nature of the crimes, this was probably the easiest of the predictions. He was a loner who used drugs and received disability checks. His mother paid the rent on his apartment. He was a college dropout.

During his questioning, Ressler asked Chase how he had selected his victims. "I go down the streets testing doors to find one that was unlocked," he explained. "If the door was locked, that means you're not welcome."

Chase was clearly paranoid. He had been released from a mental hospital only months before taking up a gun and knife. Even when he was at the hospital, the other inmates had been afraid of him and kept away. Two staff members had quit after seeing him biting the heads off birds in the garden. Chase was convinced that his blood was turning to powder and that he had to drink fresh blood to stay alive.

Despite his history of grave mental illness, the prosecution at his trial was determined not to let him get away with an insanity plea. The jury complied, and on May 8, 1978, it found Chase guilty of six counts of first-degree murder. He was sent to San Quentin's death row to await execution. While there his physical and mental health deteriorated, and he was sent to the facility for the criminally insane at Vacaville. On Christmas Eve 1980, Chase committed suicide by overdosing on anti-depressant pills he had hoarded from his daily dose.

Early in the morning of Wednesday, February 11, 1987, the mutilated body of thirty-seven-year-old Peggy Hettrick was discovered by a bicyclist on Landings Drive in Fort Collins, Colorado. At first he thought it was a mannequin lying in the field ahead of him. When he realized it really was the body of a woman, he called the police.

The coroner's autopsy showed that Hettrick had been killed by a single stab to the upper left back. Her body had subsequently been sexually mutilated.

Fifteen-year-old Timothy Masters, who lived in a house next to the field, had also seen the body when walking to school that morning. He had not gone near it, also thinking it was a mannequin left there as a prank. When the police heard from his father that Timothy had walked through the field on his way to school, they pulled him out of class and took him to the station for questioning. The lead investigator, Fort Collins detective Jim Broderick, immediately settled on Masters as his primary suspect. The boy was grilled for well over six hours with no mention of his right to an attorney. In spite of Broderick's aggressive tactics, Masters stubbornly refused to confess.

His room and school locker were searched, and a knife collection was discovered along with more than two thousand pages of his writings and drawings, many of "a pornographic and sadistic nature showing intense hostility toward women." Police also found a newspaper article about the murder. Broderick became convinced of Masters's guilt and would remain so for years despite all evidence to the contrary—there was not enough evidence for an indictment, much less a conviction.

Two hairs that were not Peggy Hettrick's were found on her body, and fingerprints that were not hers were found in her purse. They were carefully compared to Timothy Masters's, but they were not his.

The case eventually went into the cold case file, and Masters went on with his life. Five years later, in 1992, the Fort Collins police heard from a former high school classmate of Masters that when Masters and she had talked about the murder the week it happened, Masters had mentioned to her that Hettrick's nipple had been cut off. This was one of the facts that had been withheld from the public.

On the basis of this new information, the police obtained an arrest warrant. Two officers flew to Philadelphia where Masters, who had joined the navy, was stationed aboard the *USS Constitution*.

"I probably told her that," Masters admitted. "I heard it from a girl in art class."

The detectives rushed back to Fort Collins to interview this girl, who admitted that she had told Masters about the missing nipple. She had been one of a group of explorer scouts that the police had asked to help search the field for Hettrick's missing body parts. So Masters stayed in the navy.

In 1995, eight years after the murder of Peggy Hettrick, Lynn Burkhardt, a college student who was house-sitting for Dr. Richard Hammond and his family, spotted a camera lens concealed in the bathroom floor. Using a paper clip, she and a friend opened the locked door of Dr. Hammond's "spare office" on the other side of the bathroom wall. Inside they found a mass of camera and electronic equipment along with a collection of pornography, much of it pictures of women sitting on the toilet of the bathroom next door. Some of these were close-ups of breasts and genitalia. The Hammond house was one hundred yards east of where Peggy Hettrick's body was found.

When the Hammond family returned from vacation, Dr. Hammond was arrested for sexual exploitation. His wife told the police that she knew nothing of his secret room or his private hobby. She had been worried about him, however—he had recently begun collecting guns and knives.

Hammond was released on bail and checked himself into the Mountain Crest Hospital in Fort Collins for counseling. A few days later, the police were called to a motel in north Denver. Hammond had been found dead in his room. An intravenous needle containing cyanide residue protruded from his thigh. He left a suicide note that read in part, "My death should satisfy the media's thirst for blood."

At Detective Jim Broderick's direction, the Fort Collins police destroyed all the evidence in the Hammond case. The fire burned for over eight hours. Miles Moffeit, a staff writer for the *Denver Post*, had this to say about Broderick's decision: "Had Hammond been formally investigated and the evidence preserved, detectives might have been intrigued by parallels with the Hettrick case. . . . They might have searched Hammond's warehouse specifically for Hettrick's body parts. They might have tested his sex toys for DNA, as well as the knife on his belt. They might have matched his hairs with the two found on Hettrick."

But Broderick wasn't interested in Hammond; he was still focused on Masters. He had become convinced that the motive for the crime lay hidden in the mass of drawings that the police had confiscated from Masters on the day after the murder. He contacted forensic psychologist J. Reid Meloy, the author of *The Psychopathic Mind: Origins, Dynamics, and Treatment* and other books on the pathologies of criminal psychopaths. Meloy was interested in sexually motivated murders and had become an expert witness on the subject. He theorized that artwork can be a key to understanding the mind of a psychopathic killer. Not many professionals in his field agreed with him, but he was just what Broderick needed.

Meloy took a look at the pile of drawings Masters had done as a teenager and was prepared to testify that motivation had turned into action. "The killing of Ms. Hettrick translated Tim Masters' grandiose fantasy into reality," he wrote. He made this claim without interviewing or even meeting Masters. Meloy's

Forensic psychologist J. Reid Meloy interpreted this drawing by Timothy Masters to be a depiction of the homicide he had committed.

interpretation of a pencil sketch of a short man pulling another man as arrows fly toward them is bizarre.

> This is not a drawing of the crime scene as seen by Tim Masters on the morning of Feb. 11 as he went to school. This is an accurate and vivid drawing of the homicide as it is occurring. It is unlikely that Tim Masters could have inferred such criminal behavior by just viewing the corpse, unless he was an experienced forensic investigator. It is much more likely, in my opinion, that he was drawing the crime to rekindle his memory of the sexual homicide he committed the day before.

And the fact that both figures are male? And the arrows? Meloy seemed unaware of Freud's admonition that sometimes a cigar is just a cigar.

Armed with Meloy's willingness to testify about his suppositions, Broderick took the next step. On August 10, 1998, eleven years after Peggy Hettrick's murder, Tim Masters, who had settled in California after leaving the navy, was extradited on a first-degree murder charge.

The trial began in March 1999. There was no new evidence and no physical evidence of any kind connecting Tim Masters to the murder of Peggy Hettrick in a field outside Masters's house on that long-ago February day. There were only Masters's teenage drawings and a man who called himself a forensic psychologist and admitted to an obsession with sexual killings.

At the trial, the prosecution projected grossly enlarged images of Masters's drawings on the wall along with photographs of Hettrick's body. The picture of the small man dragging a man's body was one of Meloy's favorites. How did he explain the arrows? Piquerism—a sexual aberration where one is gratified by stabbing or cutting with sharp objects.

Masters was convicted and sent to prison. By a 3-2 vote the Colorado Supreme Court refused to overturn the conviction. Justice Michael Bender, writing for the dissent, said, "Most of these writings and drawings have nothing to do with this grisly murder. The sheer volume of the inadmissible evidence so overwhelmed the admissible evidence that the defendant could not have a fair trial. . . . There exists a substantial risk that the defendant was convicted not for what he did, but for who he is."

Five years into his prison sentence, Masters hired a new lawyer, Maria Liu, who quickly became convinced of his innocence. She was amazed that he had been convicted on such flimsy evidence. First, she wanted her own experts to examine the physical evidence. And the first step in this process was to have it preserved. Her motion to preserve was strenuously fought by the district attorney's office. They argued, among other things, that "There is no statutory duty to preserve evidence." In fact the two hairs found at the scene and the photographs of the unidentified fingerprints were already missing.

But Liu prevailed and took the medical examiner's photographs of Hettrick's injuries to Dr. Warren James, a Fort Collins obstetrician-gynecologist. He recognized the mutilations to the genitals as an actual surgical procedure. "Ms. Hettrick underwent a surgical procedure known as a partial vulvectomy," he said. This was not a simple procedure, he told Liu. It required a "high degree of surgical skill and high-grade surgical instruments." Powerful light would also be required, and the victim would have to be positioned with her legs apart. Tim Masters possessed neither the skill, the light, the privacy, nor the time to have done this.

On looking over the notes from the original investigation, Liu saw that even the medical examiner had called the cutting "surgical" at the time.

After a two-year court fight, Liu and her defense team finally got permission to have the remaining clothing from the crime scene sent to a lab in the Netherlands for DNA analysis. The forensic team there found no sign of Masters's DNA, but they did find a stranger's DNA on the inside of Peggy Hettrick's panties.

After twenty years in prison, Masters was released in January 2008. Several of the officials involved in the original prosecution have been investigated for unethical conduct—it was found that they had not turned over a quantity of exculpatory evidence to the defense. When asked how he felt about being largely responsible for a twenty-year miscarriage of justice, J. Reid Meloy replied, "No comment."

17 : A Burning Desire

As by air and earth and water,
so also man lives by fire.
—George R. Stewart, *Fire*

When there's a fatal fire and someone
survives, the survivor will be charged
with arson and murder.
—Gerald Hurst, Ph.D.

BY ITS VERY NATURE, arson is a difficult crime to prove. In the past it has also been a difficult crime to disprove. A perpetrator can easily start a fire when he is somewhere else—the material he requires is easily obtained, and the fire itself destroys all the evidence.

In his 1906 text *Criminal Investigation*, Hans Gross lists some of the agents used to start fires: candles, saltpeter-impregnated strips of tinder, and matches fixed to alarm clock clappers. He relates that in America:

A particularly dangerous and common method . . . is to utilize an ordinary electric bell. The sounding part of the bell is replaced by a thin, balloon-shaped glass filled with sulphuric acid. The hammer strikes on the glass and breaks it, the sulphuric acid runs into a vessel placed beneath and filled with a mixture of . . . chloric acid and sugar. This produces fire that can easily be converted into a conflagration.

According to Gross, there was once a miller's boy who started a fire in the cottage of a farmer he had a grudge against. What makes the fire noteworthy is that it was not until nine months after the boy had gone away that the farmer's house, situated next to the mill, caught fire and burned to the ground at midday, a time when everyone was working in the fields.

What the clever, evil-minded lad had done was this: He first attached a strong spring to a window in the slanted roof of the mill. This way when the window flipped open, it would throw whatever might be sitting on it onto the thatched roof of the farmer's house next door. The boy held the window shut with a string tied to the window on one end and to a ring in the wall on the other. Then he sealed the window with pitch. Next, he spread a flammable material along the path underneath the string. Then he fastened a magnifying glass (a "burning glass") to his apparatus in such a way that at a certain time of year the rays of the midday sun would be focused on his flammable material. Not until a year later was the sun in such a position that its rays were focused through the glass onto this fuel. When it caught fire, in turn it ignited the pitch and the string. As the string burned, it released the spring, which flipped open the window and tossed the burning pitch onto the farmer's roof.

According to Gross, this demonstrates "how with a little skill and ingenuity, most extraordinary things may take place."

The crime of arson is defined as "a malicious burning of property." There are more than 300,000 structural fires in the United States every year. About 75,000 of these are estimated by the FBI to be "of criminal or suspicious origin." The consensus is that 15 to 25 percent of fires involving property are arson. In times of economic distress, this figure rises. The reasons for the crime of arson are varied, and the arson investigator must first determine the arsonist's frame of mind in order to know where to look for him. Statistics show that about 90 percent of arsonists are male, and half are under the age of eighteen.

The most common motives for arson are:

—*Hate or revenge*. Setting fire to a church or synagogue or to the school or orphanage of an ethnic or religious group the arsonist dislikes is classified as a hate crime. Setting fire to the store you have just been fired from or the home of the girl who has just rejected you is an act of revenge. Almost half the arson fires set in this country are motivated by revenge.

—*Profit*. After revenge, profit is the second most common motive for arson. There are many ways to profit from burning up your own or someone else's property: collecting on insurance; getting rid of a structure you do not want and that the city or state will not permit you to tear down; getting rid of undesirable tenants; extortion; and slum clearance (so you can put up that high-rise condominium). In the Berkeley, California, hills in the 1980s, a series of grass fires was set by a part-time firefighter who needed the overtime pay.

—*Thrills*. Some people are sexually aroused by the sight of fire. Most of the pyromaniacs are men whose fire-setting began when they were quite young, possibly only eight or nine years old. The fire-setting is often compulsive—fires are repeatedly set in the same place or type of place, at the same time of day, and in the same manner.

—*Egoism*. People, sometimes even firefighters, have been known to set fires in order to appear heroic as they rescue those in danger or help put the fire out.

—*Social protest*. Some people are so caught up in a social or political cause that they blow up abortion clinics, recruiting stations, or federal buildings in the mistaken belief that they are performing a useful social service.

—*Concealment of another crime*. After a murder or robbery, what better way to get rid of the evidence than to burn it up?

Since the fire itself destroys evidence (and the longer it burns the more it destroys), an arson investigator will try to get to the scene as quickly as possible and will carefully question anyone—firefighter or civilian—who was there earlier. Some of the things the investigator will want to know are: How much smoke was there early in the fire? What colors appeared in the flames and in the smoke? Was there anything unusual about the room in the house—the presence of something unexpected or the absence of something expected?

Different materials produce flames of different colors as they burn. Smoke too will have a characteristic color. Gasoline and kerosene burn yellow or white and give off black smoke. Except for treated wood, wood generally burns yellow or red and gives off a grey smoke. If there are places on the wall where two or three pictures have been recently removed, this might be a sign that someone expected a fire.

The body of lore, rules, adages, and techniques surrounding arson and its investigation has accumulated gradually over years of practical experience by firefighters and arson investigators. But until recently much of this informed body of knowledge had not been tested scientifically. And when it was tested, much of it turned out to be incorrect.

For decades firefighters and arson investigators believed (and taught their students to believe) that:

—Fires always burn up, not down. After all, heat rises.

—Fires that burn very fast have been set using accelerants, such as kerosene or gasoline, and are therefore the results of arson.

—Accelerant-fueled fires burn hotter than normal fires.

—Separate burn holes on the floor or uneven burn patterns in one room or throughout the house indicate multiple points of origin. Accidental fires start only in one place.

—Melted copper wire or melted steel is a sign that accelerants were used.

—Bed springs and furniture springs do not collapse in a normal, or accidental, fire.

—Blistering on the walls is a sign of a hotter-than-normal fire.

—"Crazed glass"—glass that is replete with small irregular cracks—is a sign of an extra-hot fire.

None of these things is necessarily true, and some of them are quite false. But it was not until the 1990s that scientific evidence refuting the received wisdom of the ages was developed. Even today people are still being tried and convicted of arson, and murder, on the basis of faulty science.

On October 20, 1991, a wildfire in the hills of East Oakland, California, spread to nearby residential areas and burned down nearly three thousand homes before it was extinguished. Four arson investigators decided to use this tragedy as an opportunity to test some commonly held beliefs about arson fires. They found, to their surprise, that much of the conventional wisdom regarding the unmistakable signs of arson was simply not true. Fire investigator John Lentini, one of the four, has written extensively about their findings and become a vocal advocate for them. He often testifies in cases in which the prosecutors are trying to convict someone of arson by using evidence now understood to be invalid.

"Crazed glass" was supposed to be the sign of the use of accelerants. This turns out not to be so. A very hot fire will cause glass to melt but not to craze. Investigators found crazed glass in houses at the periphery of the Oakland Hills fire. All of them were houses that the fire departments had sprayed with water. It turns out that the heat from the fire alone did not craze the glass. The crazing occurred when the glass was doused with cold water when it was very hot.

So the "expert" testimony that had put people in prison over the years is wrong. Starting a fire does not cause crazed glass, but trying to put it out does.

Investigators also learned that melted copper wire was routinely found in most of the Oakland houses, and that some coils of steel (bedsprings) that appeared to have lost their tempering—which had been thought to have suffered extended high heat—actually lose tempering at a fairly low heat. And some coils which seemed to be melted, indicating an unreasonably high temperature, were actually heavily oxidized by the heat. They appeared to mimic melting unless inspected microscopically.

A phenomenon that earlier fire investigators had either missed or discounted was something called "flashover." When a fire burns in a comparatively closed space, say a room, the heat and gases rise to the ceiling and build up there until a critical point is reached. Then, in a nearly instantaneous flash, the whole room is ablaze. Often the room then burns down from the ceiling, not up from the floor. This can happen in a matter of minutes, causing a fire that burns rapidly and without the aid of accelerants or multiple points of origin.

During the night of April 7, 2003, Rose Kate Roseborough was sleeping on a couch in the living room of her home in Ashland, Ohio, when a fire started upstairs where her eleven-month-old twin daughters, Lucie and Julia Bursley, slept. She tried to reach their room and rescue them, but the smoke and heat were too intense. She ran out of the house screaming hysterically, unable to get to the telephone because it had been left upstairs in the twins' room. A neighbor called the fire department, but it was too late to save the twins, who died in the fire.

Kevin Rosser, the emergency medical technician who answered the call, claimed to notice "large-particle soot" on Roseborough's face at the scene. She was later arrested and brought to trial for arson and two counts of murder. Rosser, claiming expertise in fire examination, testified that because large-particle soot is formed only at the beginning of a fire, Roseborough must have set it herself. The defense attorney asked for a Daubert hearing (a hearing on the validity of scientific claims given in testimony), but his request was refused. In denying it, the trial judge cited a case he called *Tomlinch*, concluding that it "was easy to tell he had chopped off his thumb; he had chopped off his thumb because it was gone." Roseborough was convicted and sentenced to life without the possibility of parole.

Fortunately for Roseborough, her attorney found a real expert before her appeal. At a deposition on April 23, 2008, Gerald Hurst, an expert on fire chemistry and dynamics, testified that EMT Rosser had gotten it wrong. In fact, during a room fire the gradual depletion of oxygen and build-up of soot cause the large soot particles to form toward the *end* of the fire, not at the beginning. Hurst testified that not only did many other experts share this knowledge, but that anyone with a master's degree in chemistry could have figured it out. On January 6, 2009, the original verdict was set aside by the appeals court, and a new trial was ordered for Roseborough.

After immigrating from South Korea, Han Tak Lee, his wife Esther, and their two daughters settled in Queens, New York, where Lee ran a clothing business. Lee's older daughter, Ji Yun, suffered from mental illness and in the past had attempted suicide. On Friday, July 28, 1989, Lee and Ji Yun had a loud argument over her refusal to take her medication. Then, when Ji Yun threw a clock through a screened window the police and the pastor of the Lees' Pentecostal church were summoned to the apartment. The pastor suggested that Lee take Ji Yun to the Hebron Camp, a religious retreat run by the Korean Assembly of God Church in Stroud Township, Pennsylvania.

Ji Yun went with her father to the Hebron Camp that very day, followed later by her mother and sister. That night she had a screaming fit, and two pastors staying at the camp had to help Lee restrain her. A few hours later the Lees' cabin caught fire.

By the time the fire department arrived, the interior of the cabin was completely ablaze and flames shot out of the windows. Han Tak Lee sat on a bench a short distance away, staring silently into space. Or, as the Stroud Township Police Department put it in their report, Lee "remained complacently seated throughout the fire."

In his statement to the police, Lee said that he fell asleep after spending the evening in prayer. He was awakened by the smell of smoke and could see that his daughter's bedroom was on fire. He ran outside to see if she was there, but she was not. He then ran back inside, threw their suitcases out the door, and banged on the bathroom door. He got no response from his daughter. Finally, overcome by smoke, he ran out the back door.

Trooper Thomas Jones of the Pennsylvania State Patrol was the lead investigator for the incident. His report says that the girl's body was found in the hallway of the cabin, and that "Near the rear door of this hallway, a spill or flow pattern was observed. At this point, it was discovered that the fuel filter on the furnace had been tampered with and unscrewed, spilling fuel on the floor."

The report goes on to say that "The window on the rear wall exhibited very fine crazing of the glass and was burnt clean." Further, "spill patterns and deep char patterns" were found on the floor in various places that were "inconsistent with a normal fire."

And as for Han Tak Lee, he "remained almost emotionless and while in view of this officer made no attempts to console his wife."

On August 4, 1989, Han Tak Lee was extradited from his home in New York to answer charges of arson and murder filed against him in Stroud Township, Pennsylvania. The state assembled several expert witnesses against Lee. The first was Daniel Aston, a certified fire-protection specialist. The state based its case almost entirely on the findings in his report of December 20, 1989. Aston compared the timeline of the fire, as he reconstructed it, against a standard time-versus-temperature graph developed in 1918. Although the curve on this graph was originally meant to describe the operation of a furnace used for testing the fire resistance of various materials, it had somehow worked its way into at least one standard textbook as a means of determining whether a fire was "normal." As fire expert John Lentini put it, "Despite numerous experiments on real fires which show that this . . . time/temperature curve has no relation to reality," it was an important part of Aston's reconstruction of the fire.

Aston assigned a total weight of 7,788.28 pounds to the amount of "Class A combustible materials" in the cabin. He did this by adding the estimated weights of the cabin's contents—the sofa (60 pounds), the carpet (386.9 pounds), the mattress (32.85 pounds), and so on. He then determined the energy content of this weight of fuel, made a few incorrect assumptions about how wood burns, and created a table plotting the "normal" burn rate of such a structure against the "actual" burn rate of the cabin. He concluded that it would have required not only all of the 62 gallons of fuel missing from the cabin's fuel tank (assuming it was full) but an additional 12.2 pounds of gasoline. If Aston was right, the cabin would have been awash in flammable liquids when the fire started. And the cabin, according to Aston, burned 81.6 percent more severely than "normal."

When asked how many fires he had examined, Aston replied, "Perhaps in my career I've probably experienced some 15,000 fires that I've been called upon to determine the cause for." Now, Aston had been a part-time fire examiner for about twenty years. And if we take his number seriously, it means he investigated an average of fifteen fires a week in addition to holding down a day job designing sprinkler systems. According to John Lentini, a busy full-time fire investigator may work fifteen fires a month. The defense attorney did not do the math and failed to challenge Aston on his numbers.

Convinced as well that the fire must have been started intentionally, Lee's own attorney tried to make the case that Ji Yun had started the fire herself in an attempt to commit suicide. One assumes that if Lee had started the fire he would welcome this theory. In fact he was not at all happy with it—his daughter would not have committed suicide because it was against her religion.

"You know from common sense that this fire was started effectively," the district attorney began his summation to the jury. "Whoever did this knew what they were doing."

Unfortunately "common sense" doesn't work well when you know nothing about the subject. In fact the jury relied on the testimony of three dubious "experts":

Aston, who added in his testimony that "A shiny alligatoring pattern generally indicates a very quick fire, a dull alligatoring pattern indicates a slow, very slow fire." In fact, the alligatoring pattern on the walls indicates nothing at all about the speed of the fire.

Trooper Jones, who misinformed the jury about the significance of the crazed glass (testimony that Aston quoted and reinforced) also testified that the lack of communication between the several burn holes in the floor showed that the fire was intentionally set. In fact, this means that the roof caved in.

A chemist named Thomas Pacewicz, who testified that a "substantial quantity" of gasoline was found. On cross-examination, when asked why there was no mention of gasoline in his written report, Pacewicz said it was because "There wasn't a sufficient quantity of accelerant to confirm by our instruments." So a "substantial quantity" of gasoline was still not enough to confirm with his instruments?

Han Tak Lee was convicted of murder on September 17, 1990, and sentenced to life in prison. Several appeals have been turned down. John Lentini, who investigated the case, had this to say about it:

> The quality of the evidence presented by the Commonwealth speaks for itself. Fuel loads calculated to six significant figures, hydrocarbon "ranges" being interpreted as evidence of a mixture, furnace operating instructions being touted as normal fire behavior, and a host of other "old wives' tales" were used to convict Han Tak Lee.

Agreeing with Lentini, David M. Smith, a former bomb and arson investigator in Tucson, Arizona, said, "That's a perfect example of a system run amok."

At about four in the morning of Monday, June 30, 1986, a fire broke out in the apartment outside of Cleveland, Ohio, where two-year-old Cynthia Collins slept. Kenny Richie, a twenty-one-year-old friend of the girl's mother, was baby-sitting Cynthia while her mother was out with a friend. He was asleep and admittedly drunk in the next room. The girl was killed in the fire.

The next day Assistant State Fire Marshal Robert Cryer decided that the fire had been caused by arson—he believed that some of the burn patterns in the apartment indicated the use of accelerants. A search for empty accelerant containers turned up nothing. Nonetheless Richie was accused of felony murder on the theory that he had tried to kill his ex-girlfriend, Candy Barchet, and her new boyfriend, Mike Nichols, who lived in the apartment below. A state forensic chemist testified that gas chromatograms from the Ohio Arson Crime Laboratory showed gasoline residue on a sample from the living room carpet, and that a sample of wood from the balcony indicated the presence of paint thinner. Because the spaces between the boards had burned more than the tops of the boards, he concluded that liquid accelerant had flowed into the cracks.

William Kluge, Richie's's court-appointed attorney, with a budget of only $3,000 for expert testimony, hired L. Gregory DuBois of the firm CTL Engineering. DuBois had no experience or accreditation in arson investigation, did not interview the fire marshal, and did no testing of his own on the samples the state used to support its claims. Added to the fire marshal's opinion was the testimony of Peggy Villearreal, the next-door neighbor, who said she had heard Richie threaten to burn down the apartment building.

Despite the clear, undisputed evidence that Richie had tried desperately to get into the bedroom to save the child, Richie was convicted and sent to Ohio's death row, where he remained for the next twenty years.

In 2005 the Federal Court of Appeals for the Sixth Circuit found that "counsel's incompetent handling of the sole forensic expert in this case fell far below the wide range of acceptable professional standards." The court ordered that the State of Ohio retry Richie within ninety days or release him.

The court had learned that after the conviction, Peggy Villearreal had changed her story and said that she had never heard Richie threaten to burn anything. She also remembered that little Cynthia had a habit of playing with matches and had once set fire to the couch. But the principal factor in the court's decision was the testimony of several forensic experts that the evidence of arson cited at the trial was incorrect. Tony Café, an Australian arson expert who believed that an article of his had been misunderstood by the prosecution's experts and misquoted at the trial, filed an *amicus* brief in an attempt to put things right. He pointed out that flashover can make a fire look as though accelerants were used, and further that even without accelerants, the spaces between boards burn more than the surface because there is a good supply of air coming up through the cracks. As for the chromatograms showing gasoline on the carpet: "I cannot see any evidence whatsoever in the chromatograms that indicate the presence of an accelerant. Simply put, the chromatograms for the samples from the fire debris do not resemble the chromatograms from the standard gasoline or standard paint thinner."

And if that was not strong enough, Café added:

> I am sure that most of the world's leading forensic scientists in this field would be horrified if they saw the chromatograms used to convict Kenny Richie. If Kenny Richie were executed on the basis of this scientific evidence, then these chromatograms will become historical documents, examined by scientists all over the world and used to show just how wrong forensic evidence can be. It would be a great tragedy for the future of forensic science.

Amnesty International joined in, calling the Richie case "one of the most compelling cases of apparent innocence that human rights campaigners have ever seen."

But the state did not give up without a fight, and two years later, in 2007, the court had to repeat this same ruling. Ohio had appealed the first ruling to the Supreme Court, saying that since Richie (actually his lawyer) had not raised the question of "ineffective assistance of counsel" in earlier appeals, he could not introduce it now. The Supreme Court said that he could and sent the case back to the Sixth Circuit. There Judge R. Guy Cole Jr., wrote in the majority opinion that "the deficient performance of Richie's counsel undermines our confidence in the outcome of his trial." Again the Court gave Ohio ninety days to release Richie or give him a new trial.

This time Ohio released Richie, but not without one final dig. Richie had to plead "no contest" to a charge of child neglect and agree never to return to Putnam County, Ohio. As soon as he was released, Richie went back home to Scotland, though he later returned to the United States.

The old beliefs about how fires burn and about the telltale signs of arson are with us yet. People are still convicted of arson and even of murder based on unscientific or disproved theories. As new textbooks are written and newly trained investigators enter the field, this will gradually change. Until then people will continue to face the double horror of losing loved ones and of being held responsible for their deaths.

Select Bibliography

Alder, Ken. *The Lie Detectors*. New York, Simon and Schuster, 2007.

Bailey, William G., ed. *The Encyclopedia of Police Science*, 2nd ed. New York, Garland Publishing, 1995.

Balko, Radley. "A Case Study in Expert Testimony Gone Horribly Wrong," *Reason*, November 2007.

Block, Eugene B. *Lie Detectors: Their History and Use*. New York, David McKay, 1977.

Boos, William F. *The Poison Trail*. New York, Hale, Cushman and Flint, 1939.

Browne, Douglas G. *The Rise of Scotland Yard*. New York, G. P. Putnam's Sons, 1956.

Cole, Simon A. *Suspect Identities*. Cambridge, Mass., Harvard University Press, 2001.

DeHaan, John D. *Kirk's Fire Investigation*. Englewood Cliffs, N.J., Prentice-Hall, 1997.

Emsley, John. *The Elements of Murder*. Oxford, Oxford University Press, 2005.

Erzinçlioğlu, Zakaria. *Maggots, Murder, and Men*. New York, St. Martin's, 2000.

Faigman, David L. "Anecdotal Forensics, Phrenology, and Other Abject Lessons from the History of Science," *Hastings Law Journal*, vol. 59, no. 5 (2008).

Fisher, Jim. *Forensics Under Fire*. New Brunswick, N.J., Rutgers University Press, 2008.

Fricke, Charles W. (revised by LeRoy M. Kolbrek). *Criminal Investigation*, 6th ed. Los Angeles, Legal Book Store, 1962.

Hatcher, Julian S. *Textbook of Firearms Investigation, Identification and Evidence*. Onslow County, N.C., Small Arms Technical Publishing Company, 1935.

Hynd, Alan. *Murder, Mayhem and Mystery*. New York, A. S. Barnes, 1958.

Inbau, Fred E. *Lie Detection and Criminal Interrogation*, 2nd ed. Baltimore, Williams and Wilkins, 1948.

Kamisar, Yale. "A Look Back on a Half-Century of Teaching, Writing and Speaking About Criminal Law and Criminal Procedure," *Ohio State Journal of Criminal Law*, vol. 21 (Fall 2004).

Kurland, Michael. *How to Solve a Murder*. New York, Macmillan, 1995.

Lentini, John J. "A Calculated Arson," *Fire and Arson Investigator*, vol. 49, no. 3 (April 1999).

Lentini, John J., David M. Smith, and Richard W. Henderson. "Unconventional Wisdom: The Lessons of Oakland," *Fire and Arson Investigator*, vol. 49, no. 4 (June 1993).

McKnight, Brian E. *Law and Order in Sung China*. Cambridge, England, Cambridge University Press, 1992.

Maples, William R., and Michael Browning. *Dead Men Do Tell Tales*. New York, Doubleday, 1994.

Marsh, James. "Account of a Method of Separating Small Quantities of Arsenic from Substances with Which It May Be Mixed," *Edinburgh New Philosophical Journal*, vol. XXI (April–October 1836).

Murphy, Erin. "The Art in the Science of DNA: A Layperson's Guide to the Subjectivity Inherent in Forensic DNA Typing," *Emory Law Journal*, vol. 58 (2008).

National Research Council. *Strengthening Forensic Science in the United States: A Path Forward*. Washington, D.C., National Academies Press, 2009.

Parry, Richard. *Trial by Ice*. New York, Ballantine Books, 2001.

Paul, Philip. *Murder Under the Microscope*. London, Macdonald & Co., 1990.

Sachs, Jessica Snyder. *Corpse*. Cambridge, Mass., Perseus, 2001.

Smith, Sir Sydney. *Mostly Murder*. New York, David McKay, 1959.

Söderman, Harry. *Policeman's Lot*. New York, Funk and Wagnalls, 1956.

Thorwald, Jürgen. *The Century of the Detective*. New York, Harcourt, Brace and World, 1965.

Thorwald, Jürgen. *Crime and Science*. New York, Harcourt, Brace and World, 1967.

Tomberlin, Jeffery K., John R. Wallace, and Jason H. Byrd. "Forensic Entomology: Myths Busted!", *Forensic Magazine*, October–November 2006.

Ubelaker, Douglas, and Henry Scammell. *Bones*. New York, HarperCollins, 1992.

Wertham, Pat A. "Latent Fingerprint Evidence: Fabrication, Not Error," *The Champion*, November–December 2008.

Wilson, Colin. *The History of Murder*. New York, Carrol and Graf, 2004.

Wilson, Keith D., M.D. *Cause of Death*. Cincinnati, Writers' Digest Books, 1992.

Index

Adam, J. Collyer, 253
Aesop's Fables, 71
AFIS (automated fingerprint identification system), 89
Agrippina, 172–173
Ahern, Nancy, 210
Alder, Ken, 256
Alexander, Grand Duke, 41
Alexander VII, Pope, 173
American Academy of Forensic Sciences, 152
American Board of Forensic Odontology, 266
Amnesty International, 309
Anastasia, Grand Duchess, 40–42
The Anatomy and Physiology of the Nervous System in General . . . (Gall), 262
Anderson, Anna, 41, 228. *See also* Franziska Schanzkowska
Andrieux, Louis, 51
The Annals of Imperial Rome (Tacitus), 173
Anthropometrics, 50–57

Anthropometry, 75, 78, 80
Antistius, 159
Archimedes, 14–15
Arctic Researches and Life Among the Esquimaux (Hall), 185
Arden, Alice, 24–27
Arden, Thomas, 24–26
Arsenic, 178, 185–186, 193–194; detection of, 180–184; as popular, 179
Arson, 297, 310; and crazed glass, 301–302, 306; and flashover, 302, 309; reasons for, 298–300
The Art of Cookery (Glasse), 34
Asbury, David, 104–107
Ashton-Wolfe, Harry, 59, 62
Ashworth, Dawn, 218
Association of Firearm and Toolmark Examiners (AFTE), 153
Aston, Daniel, 305–306
Austria-Hungary, 96

A NOTE ON THE AUTHOR

Michael Kurland is perhaps best known for his "Professor Moriarty" mystery novels; he is an American Book Award nominee and two-time finalist for the Edgar. Mr. Kurland's nonfiction books include *How to Solve a Murder* and *How to Try a Murder* as well as *A Gallery of Rogues* and *The Spymaster's Handbook*. He lives in Petaluma, California.